Software Testing is a great practical help for anyone learning ... . The numerous cases, examples and exercises help to understand the software testing concepts and prepare for an ISTQB Certified Tester Foundation Level v4.0 exam. Thumbs up for this!

Kari Kakkonen, *Director of Training, Knowit Solutions Oy*

This book offers a comprehensive introduction to the basics of testing. It effectively presents key concepts through clear figures and explanations, making it easy to understand. I believe it's a valuable resource for individuals preparing for the ISTQB Foundation Level v4.0 certification exam.

Marko Rytkönen, *QA Coach at Hidden Trail*

The fifth edition of *Software Testing* is a comprehensive and clear resource for novices and professionals. It excels in explaining fundamental principles, methodologies and best practices, making it indispensable for mastering software testing. Highly recommended for certification success and excellence in the field.

Joy Aguguo, *Author, Speaker, Software Test Consultant, CEO*

This book is a veritable gold mine of information for your testing endeavours. Concise explanations, real-world examples and sample exams make even the most complex subjects easily understandable, ensuring that you ace the certification. Anyone aspiring to work as a top software tester needs to have a copy of the book – it's like having your own personal testing mentor!

Boby Jose BSc MBA MBCS, *Senior Test Manager, Sogeti UK, part of Capgemini and author of 'Test Automation: A Manager's Guide'*

A book not to ignore. This book complements ISTQB Certified Tester Foundation Level v4.0 syllabus very well, providing practical information and insight about different aspects of software testing. I especially like the fourth chapter that addresses the area of test analysis and design, as it uses so many interesting examples and many of them with images too. Highly recommended.

Kimmo Hakala, *President, Finnish Software Testing Board*

The fifth edition of *Software Testing* successfully informs the vital practice of software quality assurance and control. Providing information to help those beginning a career in software testing, the work also includes material for the more experienced practitioner and explanations of modern practices such as Agile testing.

John Burns LL.M CEng MBCS, *Information Security Risk Analyst*

SOFTWARE TESTING

BCS, THE CHARTERED INSTITUTE FOR IT

BCS, The Chartered Institute for IT, is committed to making IT good for society. We use the power of our network to bring about positive, tangible change. We champion the global IT profession and the interests of individuals, engaged in that profession, for the benefit of all.

Exchanging IT expertise and knowledge
The Institute fosters links between experts from industry, academia and business to promote new thinking, education and knowledge sharing.

Supporting practitioners
Through continuing professional development and a series of respected IT qualifications, the Institute seeks to promote professional practice tuned to the demands of business. It provides practical support and information services to its members and volunteer communities around the world.

Setting standards and frameworks
The Institute collaborates with government, industry and relevant bodies to establish good working practices, codes of conduct, skills frameworks and common standards. It also offers a range of consultancy services to employers to help them adopt best practice.

Become a member
Over 70,000 people including students, teachers, professionals and practitioners enjoy the benefits of BCS membership. These include access to an international community, invitations to a roster of local and national events, career development tools and a quarterly thought-leadership magazine. Visit www.bcs.org to find out more.

Further information
BCS, The Chartered Institute for IT,
3 Newbridge Square,
Swindon, SN1 1BY, United Kingdom.
T +44 (0) 1793 417 417
(Monday to Friday, 09:00 to 17:00 UK time)
www.bcs.org/contact

www.bcs.org/qualifications-and-certifications/certifications-for-professionals/software-testing-certifications/

http://shop.bcs.org/
publishing@bcs.uk

SOFTWARE TESTING
An ISTQB–BCS Certified Tester
Foundation Level guide (CTFL v4.0)
Fifth edition

**Geoff Thompson, Peter Morgan, Angelina Samaroo,
John Kurowski, Peter Williams and Marie Salmon
Edited by Paul Weymouth**

bcs

The
Chartered
Institute
for IT

Published by BCS Learning and Development Ltd, a wholly owned subsidiary of BCS, The Chartered Institute for IT, 3 Newbridge Square, Swindon, SN1 1BY, UK.
www.bcs.org

Paperback ISBN: 978-1-78017-6383
PDF ISBN: 978-1-78017-6390
ePUB ISBN: 978-1-78017-6406

Ebook available

British Cataloguing in Publication Data.
A CIP catalogue record for this book is available at the British Library.

Publisher's acknowledgements
Reviewers: Beth Clarke and Nicola Martin
Publisher: Ian Borthwick
Commissioning editor: Heather Wood
Production manager: Florence Leroy
Project manager: Sunrise Setting Ltd
Copy-editor: Gary Smith
Proofreader: Barbara Eastman
Indexer: David Gaskell
Cover design: Alex Wright
Cover image: istock/Peepo
Sales director: Charles Rumball
Typeset by Lapiz Digital Services, Chennai, India

CONTENTS

LIST OF FIGURES AND TABLES

AUTHORS

ABOUT THE AUTHORS

John Kurowski is a principal instructor for Expleo Group, developing and delivering courses in software testing, quality assurance, automation and Agile for over 15 years. As a member of the ISTQB Technical Advisory Group, he was a key reviewer for the alpha and beta CTFL v4.0 syllabi.

Peter Morgan is a retired test practitioner who worked as part of testing teams in a variety of industries, environments and development life cycles. Until January 2023 he was involved with writing and reviewing examination questions for the Foundation Certificate and other ISTQB exams, as well as reviewing training provider course submissions for various ISTQB qualifications, all on behalf of UK and Ireland Testing Board (UKITB). He performed both of these tasks for nearly 20 years as part of the teams responsible for those activities. These assignments began shortly after taking the Foundation Certificate in 2001 and the (old-style) Practitioner Certificate in Software Testing in 2002.

Marie Salmon is an experienced IT professional who specialises in all aspects of software testing and quality engineering. She has predominantly worked in the financial services industry and has successfully led a vast range of quality assurance services and transformation, leveraging her diverse practical and managerial experience of all levels of testing. Marie has worked with the UK and Ireland Testing Board (UKITB) for many years and is currently leading the Exams function.

Angelina Samaroo has a degree in Aeronautical Engineering from Queen Mary University of London. She is a Chartered Engineer and Fellow of the Institution of Engineering and Technology (IET). She spent 10 years working on the mission software for the Tornado ADV fighter jet. During this time, she took an interest in the career development of new engineers and led the company graduate development scheme. She then joined a small consultancy in software testing, specialising in providing training to test professionals worldwide. She now works for Pinta Education Limited and provides training in software testing, business analysis and programming.

Geoff Thompson is VP of the Qualitest BFSI sector. He has amassed a lot of experience, having worked in QA for over 30 years. He has been involved in tester certification since 1996, initially working with the BCS on the ISEB scheme as an author of the Foundation and Practitioner syllabi. He was a founder member of ISTQB, the TMMi Foundation and the UK and Ireland Testing Board, and is the immediate past Vice President of the ISTQB and Chairman of the UK and Ireland Testing Board. Geoff received the EuroSTAR

Testing Excellence Award in 2008, and the European Software Testing Awards Lifetime Achievement Award in 2015.

Paul Weymouth has over 30 years of software testing experience including test management and test consultancy in a range of government and private sector industries. He is an accredited ISTQB trainer, teaching Foundation and Advanced ISTQB certifications, and is a leading member of the UK and Ireland Testing Board (UKITB), where he runs the Technical Advisory Group, which, among other functions, reviews syllabi and associated documents for all ISTQB certifications. He is also a member of the UKITB Exams team, where he has authored and reviewed many hundreds of ISTQB Foundation and Advanced exam questions over the last decade.

Peter Williams is a self-employed contract software testing manager who has worked across a range of industries, including financial services, logistics, gambling and the public sector. He has evaluated testing processes and subsequently implemented improvements at various organisations, including introducing test management and test execution tools. He has an MSc in Computing from the Open University and was chair of the Examinations Panel for the UKITB Foundation Certificate in Software Testing.

ABOUT THE UKITB

The UK and Ireland Testing Board (UKITB) is leading the industrialisation of software testing within the UK market supporting certification and professionalisation through accreditation, education, consistency and standards adoption.

The UKITB is a not-for-profit legal entity whose roles include the support for ISTQB® and TMMi® certification schemes in the UK and Ireland and the provision of accreditation for ISTQB® trainers and training materials in the UK and Ireland via our exam partners.

The UKITB is made up of volunteers from within the UK. The directors work voluntarily on the board. All are specialists, practitioners, consultants and trainers who have experience in software testing and quality assurance positions.

The board represents a cross-section of testing companies from the UK (large companies, smaller companies and training providers) where no more than 49% of the board can be selected from any one company.

More details can be found at http://www.ukitb.org/

UKITB
UK & Ireland
Testing Board

ABBREVIATIONS

ACM	Association for Computing Machinery
ALM	application life cycle management
API	application program interface
ATDD	acceptance test-driven development
AUT	application under test
BACS	Bankers Automated Clearing Services
BDD	behaviour-driven development
BVA	boundary value analysis
CD	continuous delivery
CD	continuous deployment
CI	continuous integration
COTS	commercial off-the-shelf
CTFL	Certified Tester Foundation Level
DDD	domain-driven development
DOS	denial of service
DSL	Domain Specific Language
EP	equivalence partitioning
FDD	feature-driven development
GUI	graphical user interface
ISTQB	International Software Testing Qualifications Board
SDLC	software development life cycle
SME	subject-matter expert
SQL	Structured Query Language
TDD	test-driven development
XML	Extensible Markup Language
XP	extreme programming

PREFACE

For the previous versions of this book Brian Hambling was at the helm, coordinating all activities exceptionally well. Sadly, Brian has now retired from the UK and Ireland Testing Board and has stepped away from his managing editor role, leaving a big void that I have attempted to fill as the managing editor for the fifth edition of this book. We have also introduced some new authors for this version, Paul Weymouth, Marie Salmon and John Kurowski, who have added different and new viewpoints, which has ensured that this version remains the best supporting material for anyone wishing to take the ISTQB Foundation 4.0 exam. My fellow authors and I continue to be pleased to see how many people are finding this book so helpful in preparation for the exam.

In this fifth edition we have updated the text to include all the updates in version 4.0, including the addition of Agile content. In some instances we have kept content that is no longer part of the syllabus as we feel it adds real value to a software tester's education. We would like this new version to not only be a training aid for the exam, but also a trusted guide that software testers can continually refer to when undertaking test activities. We have refreshed many of the exercises, examples and sample questions to reflect the way the exam has evolved over the years, and for this edition we have put together a complete sample examination in Appendix A1. Appendix A3 gives commentary to explain the correct answer and why each of the distracters is incorrect. As a mock exam, the candidates can time themselves to check they are prepared for the real thing. We hope you will find this version useful and that it becomes your constant companion while testing.

INTRODUCTION

This book has been written to help potential candidates for the ISTQB CTFL examination. The book is structured to support learning of the key ideas in the syllabus quickly and efficiently for those who do not plan to attend a course, and to support structured revision for anyone preparing for the exam.

In this introductory chapter we will explain the nature and purpose of the Foundation Level and provide an insight into the way the syllabus is structured and the way the book is structured to support learning in the various syllabus areas. We offer guidance on the best way to use this book, either as a learning resource or as a revision resource.

PURPOSE OF FOUNDATION

The Foundation Certificate in Software Testing was introduced as part of the BCS Professional Certification portfolio (formerly ISEB) in 1998; since then, over 550,000 Foundation Certificates have been awarded.

The International Software Testing Qualifications Board (ISTQB) (www.istqb.org) was set up in 2001 to offer a similar certification scheme to as many countries as wished to join this international testing community. The UK was a founding member of ISTQB and, in 2005, adopted the ISTQB Foundation Certificate syllabus as the basis of examinations for the Foundation Certificate in the UK. The Foundation Certificate is an entry qualification for all ISTQB Advanced and Specialist certifications (**Figure I.1**). This book is aligned with the 2023 version of the Certified Tester Foundation Level (CTFL) syllabus, also referred to as CTFL 4.0, and includes all recent updates.

The CTFL Certificate is the first level of a hierarchy of ISTQB–BCS certificates in software testing, and leads naturally into the ISTQB Specialist and Advanced Level certifications, followed by the various ISTQB Expert Level certifications. The ISTQB website (www.istqb. org) provides details of all current Advanced, Specialist and Expert Level certifications.

The Foundation Level provides a very broad introduction to the discipline of software testing. As a result, coverage of topics is variable, with some only briefly mentioned and others studied in more detail. The arrangement of the syllabus and the required levels of understanding are explained in the next section.

The authors of the syllabus have aimed it at people with various levels of experience in testing, including those with no experience at all. This makes the certificate accessible to

Figure I.1 The ISTQB certification scheme

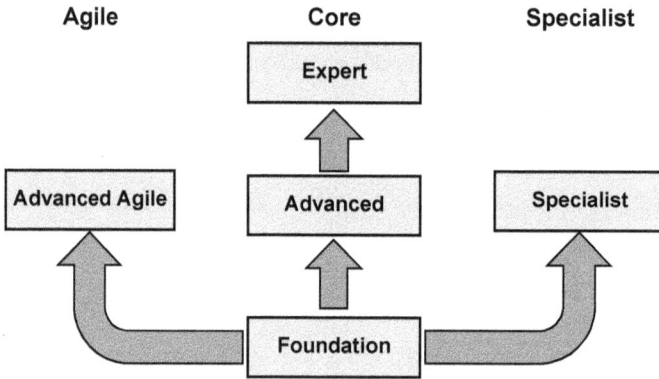

those who are or who aim to be specialist testers, but also to those who require a more general understanding of testing, such as project managers and software development managers. One aim of this qualification is to prepare certificate holders for the next level of certification, but the Foundation Level has sufficient breadth and depth of coverage to stand alone.

THE CERTIFIED TESTER FOUNDATION LEVEL SYLLABUS

Syllabus content and structure

The syllabus is broken down into six main sections, each of which has associated with it a minimum instruction time that must be included within any accredited training course:

1. Fundamentals of Testing (180 minutes)
2. Testing throughout the Software Development Life Cycle (130 minutes)
3. Static Testing (80 minutes)
4. Test Analysis and Design (390 minutes)
5. Managing the Test Activities (335 minutes)
6. Test Tools (20 minutes).

The relative timings are a reliable guide to the amount of time that should be spent studying each section of the syllabus.

Each section of the syllabus also includes a list of learning objectives that provides candidates with a guide to what they should know when they have completed their study of a section, and a guide to what can be expected to be asked in an examination.

The learning objectives can be used to check that learning or revision is adequate for each topic. In this book, which is structured around the syllabus sections, we have

presented the learning objectives for each section at the beginning of the relevant chapter, and the summary at the end of each chapter confirms how those learning objectives have been addressed.

Each topic in the syllabus has associated with it a level of understanding, represented by the legend K1, K2 or K3:

- Level of understanding K1 is associated with recall; a topic labelled K1 contains information that a candidate should be able to remember but not necessarily use or explain.
- Level of understanding K2 is associated with the ability to explain a topic or to classify information or make comparisons.
- Level of understanding K3 is associated with the ability to apply a topic in a practical setting.

The level of understanding influences the level and type of questions that can be expected to be asked about that topic in the examination. More detail about the question style and about the examination is given in **Chapter 7**.

Example questions, written to the level and in the formats used in the examination, are included within each chapter to provide generous examination practice. There is one complete practice examination paper in **Appendix A**, with answers and an explanation of the correct response to each question.

Syllabus updates and changes

The syllabus was last updated in 2023, and the book is in line with the changes and additions introduced in that version (also referred to as CTFL 4.0). This is a major update to the syllabus, which effectively merges key topics that reflect modern testing practices from the 2018 Software Testing Foundation syllabus and the 2014 Software Testing Level Agile Tester (CTFL-AT) syllabus. Additional topics were added, such as DevOps and shift-left. The effect of this amalgamation is that it is no longer necessary for candidates to obtain the CTFL-AT qualification in addition to the CTFL qualification in order to study for the ISTQB Advanced Agile qualifications.

Throughout this book you will find references to different testing standards that form reference points for parts of the ISTQB Software Testing Foundation syllabus:

- ISO/IEC/IEEE 29119
- ISO/IEC 25010
- ISO/IEC 20246.

ISO/IEC/IEEE 29119 *Software Testing* is an internationally agreed set of standards for software testing that can be used within any software development life cycle (SDLC) or organisation. There are currently five standards within ISO/IEC/IEEE 29119:

- ISO/IEC/IEEE 29119-1: *General concepts* (published January 2022) facilitates understanding and use of all other standards and the vocabulary used within the 29119 series.

- ISO/IEC/IEEE 29119-2: *Test processes* (published October 2021) defines a generic process model for software testing that can be used within any SDLC.

- ISO/IEC/IEEE 29119-3: *Test documentation* (published October 2021) defines templates for test documentation that cover the entire software testing life cycle.

- ISO/IEC/IEEE 29119-4: *Test techniques* (published October 2021) defines software test design techniques (also known as test case design techniques or test methods).

- ISO/IEC/IEEE 29119-5: *Keyword-driven testing* (published November 2016) defines an international standard for supporting keyword-driven testing.

RELATIONSHIP OF THE BOOK TO THE SYLLABUS

The book is structured in chapters that mirror the sections of the syllabus so that you can work your way through the whole syllabus or select topics that are of particular interest or concern. The structure enables you to go straight to the place you need, with confidence either that what you need to know will be covered there and nowhere else, or that relevant cross-references will be provided.

Each chapter incorporates the learning objectives from the syllabus and identifies the required level of understanding for each topic. Each chapter also includes examples of typical examination questions to enable you to assess your current knowledge of a topic before you read the chapter. There are further questions at the end of each chapter to provide practice in answering typical examination questions. Topics requiring a K3 level of understanding are presented with worked examples as a model for the level of application expected from real examination questions. Answers are provided for all questions, and the rationale for the correct answer is discussed for all practice questions.

A final chapter explains the Foundation Level examination strategy and provides guidance on how to prepare for the examination and how to manage the examination experience to maximise your performance.

HOW TO GET THE BEST OUT OF THIS BOOK

This book is designed for use by different groups of people. If you are using the book as an alternative to attending an accredited course, you will probably find the first method described below to be of greatest value. If you are using the book as a revision aid, you may find the second approach more appropriate. In either case, you would be well advised to acquire a copy of the syllabus and a copy of the sample examination papers (both available free from www.istqb.org) as reference documents, though neither is essential and the book stands alone as a learning and revision aid.

Using the book as a learning aid

For those of you using the book as an alternative to attending an accredited course, the first step is to familiarise yourself with the syllabus structure and content by skim-reading the opening sections of each chapter, where the learning objectives are identified for each topic. You may then find it helpful to turn to **Chapter 7** and become familiar with the structure of the examination and the types and levels of questions that you can expect. From here you can work through each of the six main chapters in any sequence before returning to **Chapter 7** to remind yourself of the main elements of the examination.

For each chapter begin by attempting the self-assessment questions at the beginning to get initial confirmation of your level of confidence in the topics covered by that chapter. This may help you to prioritise how you spend your time. Work first through the chapters where your knowledge is weakest, attempting all the exercises and following through all the worked examples.

Read carefully through the chapters where your knowledge is less weak, but still not good enough to pass the exam. You can be more selective with exercises and examples here, but make sure you attempt the practice questions at the ends of the chapters. For the areas where you feel strong you can use the chapter for revision, but remember to attempt the practice questions to confirm positively your initial assessment of your level of knowledge.

Each chapter contains 'checks of understanding' between sections so that you can determine whether you have picked up the key points from that section. These are important; if you find you cannot answer them, you would be wise to go back over the material to revise anything that you did not really absorb at the first read. There is also a summary section at the end of each chapter that reiterates the learning objectives, so reading the first and last sections of a chapter will help you to understand how your current level of knowledge relates to the level required to pass the examination. The best confirmation of this is to attempt questions at the appropriate K level for each topic; these are provided throughout the book and there is a complete mock examination at the end so that you can really test whether you are ready for the exam.

Using the book as a revision aid

If you are using this book for final revision, perhaps after completing an accredited course, you might like to begin by using a selection of the example questions at the end of each chapter as a 'revision mock examination'. **Appendix A1** contains one complete mock exam, with all the answers in **Appendix A2**, which will provide some experience of answering typical questions under the same time pressures that you will experience in the real examination. This will provide you with a fairly reliable guide to your current state of readiness to take the real examination. You can also discover which areas most need revision from your performance in the mock exam, and this will guide you as you plan your revision. There is a complete question-by-question commentary on the mock exam in **Appendix A3** so that you can identify why you got any questions wrong.

Revise first where you feel weakest. You can use the opening sections of each chapter, containing the learning objectives and the self-assessment questions, together with the

'Check of understanding' provided throughout the chapters, and the chapter summary at the end of each chapter, to refine further your awareness of your weaknesses. From here you can target your studies very accurately. Remember that every K3 topic will have at least one worked example and some exercises to help you build your confidence before tackling questions at the level set in the real examination.

You can get final confirmation of your readiness to sit the real examination by taking the ISTQB sample examination papers available from BCS.

1 THE FUNDAMENTALS OF TESTING

Peter Morgan

COVID-19 was a game-changer in the UK and many other countries. Ways of doing ordinary things that you used to do in-person (like shopping for everyday items) you may now do online. Whenever and wherever you buy something, you expect it to match the description you were given. An ex-showroom new car should not have a large scratch down a door panel, it should have the correct number of wheels and come with an owner's handbook. The vehicle should have the correct engine size, and the performance in areas such as fuel consumption, battery charging time and maximum speed should match published figures. In short, a level of expectation is set by brochures, by your experience of sitting in the driving seat and probably by a test drive. If your expectations are not met, you will feel justifiably aggrieved.

New software installations often fall short of expectations. Why is this? There is no single cause that can be rectified to solve the problem, but one important contributing factor is the inadequacy of the testing to which software applications are exposed.

Software testing **can be simple**, but it is a discipline that is seldom applied with anything approaching the necessary rigour to provide confidence in the delivered software. Software testing is costly in human effort or in the technology that can multiply the effect of human effort, and it is seldom implemented at a level that will provide assurance that software will operate effectively, efficiently or even correctly.

This book explores the fundamentals of this important discipline to provide a basis on which a practical and cost-effective software testing regime can be constructed.

INTRODUCTION

This opening chapter has three important objectives:

1. It introduces you to the fundamental ideas that underpin the discipline of software testing; this will involve the use and explanation of new terminology.

2. It establishes the structure used throughout the book to help you to use the book as a learning and revision aid.

3. It points forward to the content of later chapters.

The key ideas of software testing are applicable irrespective of the software involved and any particular development methodology (waterfall, Agile, etc.). Software development methodologies are discussed in detail in **Chapter 2**.

We begin by defining what we expect you to get from reading this chapter. The learning objectives below are based on those defined in the Software Foundation Certificate syllabus, so you need to ensure that you have achieved all of these objectives before attempting the examination.

Learning objectives

The learning objectives for this chapter are listed below. You can confirm that you have achieved these by using the self-assessment questions immediately following the learning objectives, the 'Check of understanding' boxes distributed throughout the text and the example examination questions provided at the end of the chapter. The chapter summary will remind you of the key ideas.

Each learning objective is allocated a K number to represent the level of understanding required; see the **Introduction** (pp. 2–3) for an explanation of K numbers.

What is testing?

- FL-1.1.1 (K1) Identify typical test objectives.
- FL-1.1.2 (K2) Differentiate testing from debugging.

Why is testing necessary?

- FL-1.2.1 (K2) Exemplify why testing is necessary.
- FL-1.2.2 (K1) Recall the relation between testing and quality assurance.
- FL-1.2.3 (K2) Distinguish between root cause, error, defect, and failure.

Testing principles

- FL-1.3.1 (K2) Explain the seven testing principles.

Test activities, testware and test roles

- FL-1.4.1 (K2) Summarise the different test activities and tasks.
- FL-1.4.2 (K2) Explain the impact of context on the test process.
- FL-1.4.3 (K2) Differentiate the testware that supports the test activities.
- FL-1.4.4 (K2) Explain the value of maintaining traceability.
- FL-1.4.5 (K2) Compare the different roles in testing.

Essential skills and good practices in testing

- FL-1.5.1 (K2) Give examples of the generic skills required for testing.
- FL-1.5.2 (K1) Recall the advantages of the whole team approach.
- FL-1.5.3 (K2) Distinguish the benefits and drawbacks of independence of testing.

Self-assessment questions

The following questions have been designed to enable you to check your current level of understanding for the topics in this chapter. The answers are at the end of the chapter.

Question SA1 (K2)
Which of the following correctly describes the interdependence between an error, a defect and a failure?

    A. An error causes a failure that can lead to a defect.
    B. A defect causes a failure that can lead to an error.
    C. An error causes a defect that can lead to a failure.
    D. A failure causes an error that can lead to a defect.

Question SA2 (K2)
An online site where goods can be bought and sold has been implemented. Which one of the following could be the root cause of a defect?

    A. Customers complain that the time taken to move to the 'payments' screen is too long.
    B. The lead business analyst was not familiar with all possible permutations of customer actions.
    C. Multiple customers can buy the **one** item that is for sale.
    D. There is no project-wide defect-tracking system in use.

Question SA3 (K2)
Which of the following illustrates the principle of defect clustering?

    A. Testing everything in most cases is not possible.
    B. Even if no defects are found, testing cannot show correctness.
    C. It is incorrect to assume that finding and fixing a large number of defects will ensure the success of a system.
    D. A small subset of the code will usually contain most of the defects discovered during the testing phases.

WHY SOFTWARE FAILS

Examples of software failure are depressingly common. Here are some you may recognise:

- A data breach at T-Mobile® led to the confidential records of 50 million customers being compromised, including sensitive information (phone numbers and bank account details).

- After successful test flights and air worthiness accreditation, problems arose in the manufacture of the Airbus A380™ aircraft. Assembly of the large subparts into the completed aircraft revealed enormous cabling and wiring problems. The wiring of large subparts could not be joined together. It has been estimated that the direct or indirect costs of rectification were $6.1 billion. (Note: this problem was quickly fixed and the aircraft entered into service within 18 months of the cabling difficulties being identified.)

- On 3 May 2021, TikTok users found they had 0 (zero) followers of their account. These things matter to a lot of people.

- When the UK Government introduced online filing of tax returns, a user could sometimes see the amount that a previous user earned. This was regardless of the physical location of the two applicants.

- One of the new features of the *Call of Duty: Warzone*™ game had to be pulled the same day that the product was launched as it produced unexpected – and undesirable – consequences.

- In November 2005, information on the UK's top 10 wanted criminals was displayed on a website. The publication of this information was described in newspapers and on morning radio and television and, as a result, many people attempted to access the site. The performance of the website proved inadequate under this load and it had to be taken offline. The publicity created performance peaks beyond the capacity of the website.

- A catastrophic disruption of service in the Federal Aviation Authority (FAA) computer service in early 2023 resulted in services across the whole of the USA being grounded for parts of 11 January 2023.

- A new smartphone mapping application was introduced in September 2012. Among many other problems, a museum was incorrectly located in the middle of a river, and Sweden's second city, Gothenburg, seemed to have disappeared from at least one map.

Perhaps these examples are not the same as the notorious final payment demands for 'zero pounds and zero pence' of some utilities customers in the 1970s. But what is it that still makes them so startling? Is it a sense that something fairly obvious was missed? Is it the feeling that, expensive and important as they were, the systems were allowed to enter service before they were ready? Do you think these systems were adequately tested? Obviously they were not, but in this book we want to explore why this was the case and why these kinds of failures continue to plague us.

To understand what is going on we need to start at the beginning, with the people who design systems. Do they make mistakes? Of course they do. People make mistakes because they are fallible, but there are also many pressures that make mistakes more likely. Pressures such as deadlines, complexity of systems and organisations, and changing technology all bear down on designers of systems and increase the likelihood of defects in specifications, in designs and in software code. Errors are where major system failures usually begin. An error is best thought of as 'invisible', or better perhaps as intangible – you cannot touch it. It is an incorrect thought, a wrong assumption or a thing that is forgotten or not considered. Only when something is written down can it become a 'fault' or a defect. So, an incorrect choice can lead to a document with a defect in it, which, if it is used to specify a component, can result in the component being faulty and maybe exhibiting incorrect behaviour. If this faulty component is built into a system, the system may fail. While failure is not guaranteed, it is likely that errors in the thought processes as specifications are produced will lead to faulty components, and faulty components will cause system failure.

An error (or mistake) leads to a defect (or fault or bug), which can cause an observed failure (**Figure 1.1**).

Figure 1.1 Effect of an error

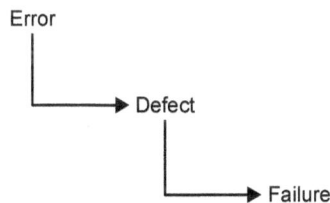

There are other reasons why systems fail. Environmental conditions such as the presence of radiation, magnetism, electronic fields or pollution can affect the operation of hardware and firmware and lead to system failure.

It is worth pointing out that not every apparent failure is a real failure – something appears to be a failure in the software, but the observed behaviour is correct. Perhaps the tester who created the test misunderstood what should happen in the precise circumstances. When an apparent failure in a test is actually the application or system performing correctly, this is termed a **false positive**. Conversely, a **false negative** is where there is a real failure but this is not identified as such and the test is seen as correct.

If we want to avoid failure, we must either avoid errors and faults or find them and rectify them. Testing can contribute to both avoidance and identification, as we will see when we look at the test process in a little more detail. One thing is clear: if we wish to identify errors through testing we need to begin testing as soon as we begin making errors – right at the beginning of the development process – and we need to continue testing until we are confident that there will be no serious system failures – right at the end of the development process.

Before we move on, let us just remind ourselves of the importance of what we are considering. Incorrect software can harm:

- people (e.g. by causing an aircraft crash in which people die, or by causing a hospital life support system to fail);
- companies (e.g. by causing incorrect billing, which results in the company losing money);
- the environment (e.g. by releasing chemicals or radiation into the atmosphere).

Software failures can sometimes cause all three of these at once. The scenario of a train carrying nuclear waste being involved in a crash has been explored to help build public confidence in the safety of transporting nuclear waste by train. A failure of the train's on-board systems or of the signalling system that controls the train's movements could lead to catastrophic results. This may not be likely (we hope it is not) but it is a possibility that could be linked with software failure. Software failures, then, can lead to:

- loss of money;
- loss of time;
- loss of business reputation;
- injury;
- death.

KEEPING SOFTWARE UNDER CONTROL

With all of the examples we have seen so far, what common themes can we identify? There may be several themes that we could draw out of the examples, but one theme is clear: either insufficient testing or the wrong type of testing was done. More and better software testing seems a reasonable aim, but that aim is not quite as simple to achieve as we might expect.

Exhaustive testing of complex systems is not possible

The launch of the smartphone mapping app occurred at the same time as a new phone hardware platform – the new app was only available on the new hardware for what many would recognise as the market leader at that time. The product launch received extensive worldwide coverage, with the mapping app given a prominent place in launch publicity. In a matter of hours there were tens of thousands of users, and numbers quickly grew, with many people wanting to see their location in the new app and see how this compared with (for example) Google Maps. Each phone user was an expert in their own location – after all, they lived there – and 'test cases' were generated showing that problems existed.

If every possible test had been run, problems would have been detected and rectified prior to the product launch. However, if every test had been run, the testing would still be running now (over a decade later), and the product launch would never have taken place. This illustrates one of the general principles of software testing, which are explained below. With large and complex systems it will never be possible to test everything exhaustively; in fact, it is impossible to test even moderately complex systems exhaustively.

For the mapping app it would be unhelpful to say that not enough testing was done; for this particular project, and for many others of similar complexity, that would certainly always be the case. Here the problem was that the right sort of testing was not done because the problems had not been detected.

Testing and risk

Risk is inherent in all software development. The system may not work or the project to build it may not be completed on time, for example. These uncertainties become more significant as the system complexity and the implications of failure increase. Intuitively, we would expect to test an automatic flight control system more than we would test a video game system. Why? Because the risk is greater. There is a greater probability of failure in the more complex system and the impact of failure is also greater. What we test, and how much we test it, must be related in some way to the risk. Greater risk implies more and better testing.

There is much more on risk and risk management in **Chapter 5**.

Testing and quality

Quality is notoriously hard to define. If a system meets its users' requirements, that constitutes a good starting point. In the examples we looked at earlier, the online tax returns system had an obvious functional weakness in allowing one user to view another user's details. While the user community for such a system is potentially large and disparate, it is hard to imagine any user that would find that situation anything other than unacceptable. In the top 10 criminals example, the problem was slightly different. There was no failure of functionality in this case; the system was simply swamped by requests for access. This is an example of a non-functional failure, in that the system was not able to deliver its services to its users because it was not designed to handle the peak load that materialised after radio and TV coverage.

The problems with the Airbus A380 aircraft is an interesting story, because although completed subparts could not be brought together to build an entire aircraft, each of the subparts was 'correctly manufactured'. Aircraft are increasingly sophisticated, and the A380 aircraft has approximately 530 km of cables, 100,000 wires and 40,300 connectors. Software is used both to design the aircraft and in its manufacture. However, the large subparts were made in two different countries, with different versions of the software in each manufacturing base. When Airbus was bringing together two parts of the aircraft, the different software meant that the wiring on one part did not match the wiring on the other. The cables could not meet up without being changed. Testing may have taken place, but it did not test something as straightforward as the versions of the design software and whether they were compatible (in this case they were plainly not). Each large subpart was built according to its own version of the CATIA (Computer Aided Three-Dimensional Interactive Application) software. The result did not provide an aircraft that could fly.

Of course, the software development process, like any other process, must balance competing demands for resources. If we need to deliver a system faster (i.e. in less time), it will usually cost more. The items at the corners (or vertices) of the triangle of resources in **Figure 1.2** are time, money and quality. These three items affect one another and also influence the features that are or are not included in the delivered software.

Figure 1.2 Resources triangle

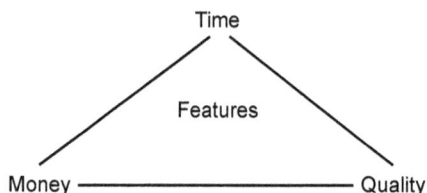

One role for testing is to ensure that key functional and non-functional requirements are examined before the system enters service, and any defects are reported to the development team for rectification. Testing cannot directly remove defects, nor can it directly enhance quality. Reporting defects makes their removal possible and so contributes to the enhanced quality of the system. In addition, the systematic coverage of a software product in testing allows at least some aspects of the quality of the software to be measured. Testing is one component in overall quality assurance that seeks to ensure that systems enter service without defects that can lead to serious failures.

Deciding when 'enough is enough'

How much testing is enough, and how do we decide when to stop testing?

We have so far decided that we cannot test everything, even if we would wish to. We also know that every system is subject to risk of one kind or another and that there is a level of quality that is acceptable for a given system. These are the factors we use to decide how much testing to do.

The most important aspect of achieving an acceptable result from a finite and limited amount of testing is prioritisation. Do the most important tests first so that at any time you can be certain that the tests that have been done are more important than the ones still to be done. Even if the testing activity is cut in half, it will still be true that the most important testing has been done. The most important tests will be those that test the most important aspects of the system: they will test the most important functions as defined by the users or sponsors of the system, and the most important non-functional behaviour, and they will address the most significant risks.

The next most important aspect is setting criteria that will give you an objective measure of whether it is safe to stop testing, so that time and all the other pressures do not confuse the outcome. These criteria, usually known as acceptance criteria, set the standards for the testing activity by defining areas such as how much of the software is to be tested (this is covered in more detail in **Chapter 4**) and what levels of defects can be tolerated in a delivered product (covered in more detail in **Chapter 5**).

Priorities and completion criteria provide a basis for planning (covered in **Chapters 2 and 5**), but the triangle of resources in **Figure 1.2** still applies. In the end, we may have to compromise in terms of the desired level of quality and risk, but our approach ensures that we can still determine how much testing is required to achieve the agreed levels and we can still be certain that any reduction in the time or effort available for testing will not affect the balance – the most important tests will be those that have already been done, whenever we stop.

CHECK OF UNDERSTANDING

1. Describe the interaction between errors, defects and failures.
2. Software failures can cause losses. Give three consequences of software failures.
3. What are the vertices of the 'resources triangle'?

WHAT TESTING IS AND WHAT TESTING DOES

So far, we have worked with an intuitive idea of what testing is. We have recognised that it is an activity used to reduce risk and improve quality by finding defects, which is all true. There are indeed many different definitions of 'testing' when applied to software. Here is one that we have found useful – but don't worry: you are not expected to remember it.

Testing is the systematic and methodical examination of a work product using a variety of techniques, with the express intention of attempting to show that it does not fulfil its desired or intended purpose. This is undertaken in an environment that represents most nearly that which will be used in live operation.

Like other definitions of testing, it is not perfect. But it does point the way to material that will be covered in later chapters of this book, including:

- systematic – it has to be planned (**Chapter 5**);
- methodical – it is a process (throughout, but initially later in this chapter);
- work product – it is not just code that is examined (**Chapter 3**);
- variety of techniques (**Chapter 4**).

Any definition of testing has limitations. The definition depends upon the aim(s) or goals of that testing, or indeed the testing of that application. Not all testing has the same aim, and the aim of testing can vary through the software development life cycle (SDLC).

So, what does testing aim to do? Here are some of the principle objectives of testing:

- to examine work products (note it is 'work product'; as well as code, this can include documentation such as requirements, user stories and the overall design, among other items);
- to check if all the requirements have been satisfied (system verification – see below);
- to see whether the item under test is complete, and works as the users and other stakeholders expect (system validation – see below);
- to instil confidence in the quality of the item under test;
- to find failures and defects and prevent these reaching the production version of the software – often this is the first item that people will list;
- to give sufficient information to enable decision makers to make decisions, perhaps about whether the software product is suitable for release;
- to reduce the level of risk of inadequate software quality (e.g. previously undetected failures occurring in operation);
- to comply with contractual, legal or regulatory requirements or standards. Some standards require, for example, 80 per cent decision coverage to be shown by testing.

Verification: Checking that the development process is undertaken correctly, that no requirements have been lost or misinterpreted at (say) the coding stage. Often summed up by 'Are we building the system right?'

Validation: Determining whether the system achieves what is wanted – which can sometimes be a challenge when there are several groups of users (clerk, supervisor, retail manager, etc.). 'Are we building the right system?'

Testing can therefore be a multifaceted activity. We need to understand a little more about how software testing works in practice before we can think about how to implement effective testing.

Before that, though, we will sum up what testing is by looking at a few of the key terms or definitions that are relevant to this opening chapter. Remember, the keywords that are given in the syllabus for this section of material can be used in examination questions, where a clear understanding of the meaning of the terms is required.

Testing relies upon an understanding of what the system, application or utility is meant to achieve. This is usually, but not always, written down, perhaps in a requirements document, in a description of the house style of web applications in the company or in the interface definitions that are required for external systems. This body of knowledge is termed the **test basis**; testing has the test basis as a key starting point for what to test and how to test.

Within the test basis there are descriptions of what should happen under certain circumstances – 'if the username and password combination is incorrect, display an error message and request that the operator tries again; if the details are incorrect on the third attempt, lock the user account'. Such descriptions are called **test conditions** – what will happen as a consequence of previous choices or actions. Test conditions give the circumstances, but not usually the values required to run a test.

The username or password may indeed be incorrect, but the specific values to be used are given in a **test case**. A test case is derived from one or more test conditions and describes in some detail what is required to enter into the application or system, and most importantly what the expected outcome is, so that we know whether the test is successful or not (has 'passed' or 'failed'). In many instances there are preconditions that must be in place before one or more particular test cases can be run and actions that are to be done when the test case or cases have been run. This is called a **test procedure**. Here are the definitions of these four terms, taken from the ISTQB glossary:

Test basis: The body of knowledge used as the basis for test analysis and design.

Test condition: A testable aspect of a component or system identified as a basis for testing.

Test case: A set of preconditions, inputs, actions (where applicable), expected results and postconditions, developed based on test conditions.

Test procedure: A sequence of test cases in execution order, and any associated actions that may be required to set up the initial preconditions and any wrap-up activities post-execution.

Static testing and dynamic testing

Static testing is the term used for testing when the code is not exercised. This may sound strange, but remember that failures often begin with a human error, namely a wrong way of thinking or an incorrect assumption (an error) when producing a document such as a specification (which will then have a defect in it). We need to test as early as possible because errors are much cheaper to fix than defects or failures (as you will see later). We discussed earlier that errors are intangible, but the earlier we find something that is incorrect, the easier (and cheaper) it is to fix. That is why testing should start as early as possible (another basic principle explained in more detail later in this chapter). Static testing involves techniques such as reviews, which can be effective in preventing defects in the resulting software – for example, by removing ambiguities, omissions and faults from specification documents. This is a topic in its own right and is covered in detail in **Chapter 3**. Dynamic testing is the kind that exercises the program under test with some test data, so we speak of test execution in this context. The discipline of software testing encompasses both static and dynamic testing.

These activities are typically undertaken at different times in the SDLC. To illustrate this, we will briefly consider the V model development process (**Figure 1.3**), although this is introduced only to illustrate this point at this stage, being described in detail in

Chapter 2. **Static testing** is undertaken in the left side of this diagram (earlier in time), and **dynamic testing** to the right (later in time).

Figure 1.3 The place of static testing versus dynamic testing

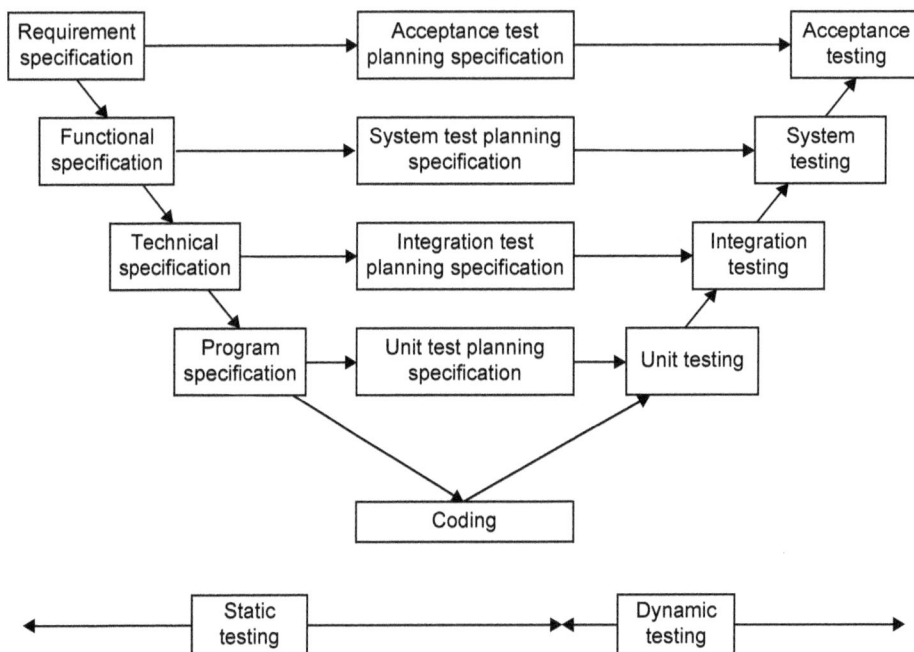

```
┌──────────────┐         ┌────────────────────┐         ┌──────────────┐
│ Requirement  │────────▶│  Acceptance test   │────────▶│  Acceptance  │
│ specification│         │planning specification│        │   testing    │
└──────────────┘         └────────────────────┘         └──────────────┘
      │
      ▼
 ┌──────────────┐        ┌────────────────────┐         ┌──────────────┐
 │  Functional  │───────▶│  System test planning│──────▶│   System     │
 │ specification│        │    specification    │        │   testing     │
 └──────────────┘        └────────────────────┘         └──────────────┘
        │
        ▼
   ┌──────────────┐      ┌────────────────────┐        ┌──────────────┐
   │  Technical   │─────▶│  Integration test  │───────▶│ Integration  │
   │ specification│      │planning specification│       │   testing    │
   └──────────────┘      └────────────────────┘        └──────────────┘
          │
          ▼
     ┌──────────────┐    ┌────────────────────┐       ┌──────────────┐
     │   Program    │───▶│ Unit test planning │──────▶│ Unit testing │
     │ specification│    │    specification    │       │              │
     └──────────────┘    └────────────────────┘       └──────────────┘

                         ┌────────────────────┐
                         │       Coding        │
                         └────────────────────┘
```

```
◀──────────┌──────────┐──────────────────────▶◀─┌──────────┐──────────▶
           │  Static  │                          │ Dynamic  │
           │ testing  │                          │ testing  │
           └──────────┘                          └──────────┘
```

Testing and debugging

Testing and debugging are different kinds of activities, both of which are very important. Testing is about finding problems (defects or failures), whereas debugging concerns the resolution of these. Testing can show where there are unintended consequences (failures), and these are usually reported on defect reports (see **Chapter 5**).

Debugging is the process that developers go through to identify the cause of failures – in other words, defects in code – and undertake corrections. Ideally, some check of the correction is made, but this may not extend to checking that other areas of the system have not been inadvertently affected by the correction.

Testing is a systematic exploration of a component or system with the main aim of finding and reporting defects. Testing does not include correction of defects – these are passed to the developer to correct. Testing does, however, ensure that changes and corrections are checked for their effect on other parts of the component or system. Where a defect is found by static testing, debugging action is concerned with removing the defect, whereas when dynamic testing triggers a failure, debugging involves finding causes, analysing them and removing these. This can involve reproducing any failures.

Replicating a problem is not necessary (or indeed possible) in static testing, as a failure has not occurred.

Usually, developers will have undertaken some initial testing (and taken appropriate corrective action as necessary – debugging) to raise the level of quality of the component or system to one that is worth testing; that is, a level that is sufficiently robust to enable rigorous testing to be performed. Debugging does not give confidence that the component or system meets its requirements completely. Testing makes a rigorous examination of the behaviour of a component or system and reports all defects found to the development team for investigation and possible correction. Testing then repeats enough tests to ensure that defect corrections have been effective. So, both are needed to achieve a quality result. The sequence of actions is test–debug–test, with the last test(s) including both confirmation testing and regression testing.

While it is generally true that testers test and developers undertake debugging actions, this is not always so clear-cut. In some software development methodologies (typically Agile development, but not restricted to this), testers are routinely involved in both component testing and debugging activities.

Defects and root causes

We talked earlier about the differences and interconnections between an error, a defect and a failure. There is a similar relationship between the root cause of a defect and the defect. Why a defect took place (the root cause) may have little relationship to the defect (what is wrong). Some organisations routinely undertake root cause analysis of defects, with the aim of preventing similar problems happening in the future. If we can identify the root cause of a defect, steps can be taken to try to prevent similar defects with the same root cause in future. This could be further training, ensuring that inexperienced developers buddy-up with someone who has greater understanding of (for example) the development language. However, be aware that similar or even identical defects can have completely different root causes.

The defect is what is wrong (an incorrect calculation of interest rates), the effect is how this appears (an angry customer who is charged too much) and the root cause is why the defect came about (an incorrect understanding by the business lead of how long-term interest rates are to be calculated). In considering defects, it is important to distinguish between the root cause (the 'why') and the effects (the 'what happened') of a defect.

Testing and quality assurance

Testing is part of quality control, which is included in quality management (itself comprising quality control and quality assurance) (**Figure 1.4**).

As we shall see, quality control encompasses more than testing. It is a project activity focused on seeing if the desired level of quality is being achieved (and if not, doing something about it). In contrast to this, quality assurance is about making sure that processes are undertaken correctly. It is usually an organisational activity and is a process-oriented, preventive approach that focuses on the implementation of processes and, if applicable, the improvement of these processes. If processes are carried out correctly, there is a greater likelihood that the end product will be better.

Figure 1.4 Quality management includes both quality assurance and quality control

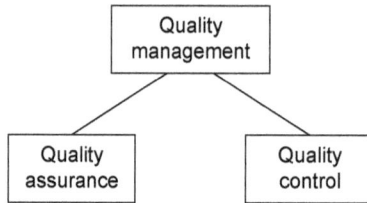

```
         ┌──────────────┐
         │   Quality    │
         │  management  │
         └──────────────┘
            ╱        ╲
   ┌──────────────┐  ┌──────────────┐
   │   Quality    │  │   Quality    │
   │  assurance   │  │   control    │
   └──────────────┘  └──────────────┘
```

Testing is part of quality control (checking the quality of something – in this case, the software under test), but it does not necessarily have a part to play in courses of action if the quality does not match the desired level. It gives an indication of how well the development and test processes are performing. Of course, checking quality can be undertaken in ways other than testing (e.g. through questionnaires). Testing does not in itself improve quality, but it does measure the quality and highlights areas where there may be room for improvement. We say 'may be' as a problem reported by those testing may not require corrective action in the software, but needs the so-called failed test to be changed.

Testing as a process

We have already seen that there is much more to testing than test execution. Before test execution, there is some preparatory work to do to design the tests and set them up; after test execution, there is some work needed to record the results and check whether the tests are complete. Even more important than this is deciding what we are trying to achieve with the testing and setting clear objectives for each test. A test designed to give confidence that a program functions according to its specification, for example, will be quite different from one designed to find as many defects as possible. We define a test process to ensure that we do not miss critical steps and that we do things in the right order. We will return to this important topic later, when we explain a generalised test process in detail.

Testing as a set of techniques

The final challenge is to ensure that the testing we do is effective testing. It might seem paradoxical, but a good test is one that finds a defect if there is one present. A test that finds no defect has consumed resources but not necessarily added value; a test that finds a defect has created an opportunity to improve the quality of the product. How do we design tests that find defects? We actually do two things to maximise the effectiveness of the tests. First, we use well-proven test design techniques; a selection of the most important of these is explained in detail in **Chapter 4**. The techniques are all based on certain testing principles that have been discovered and documented over the years, and these principles are the second mechanism we use to ensure that tests are effective. Even when we cannot apply rigorous test design for some reason (such as time pressures), we can still apply the general principles to guide our testing. We turn to these next.

TESTING PRINCIPLES

Testing is a very complex activity, and the software problems described earlier highlight that it can be difficult to do well. We now describe some general testing principles that help testers, principles that have been developed over the years from a variety of sources. These are not all obvious, but their purpose is to guide testers and prevent the types of problems described previously. Testers use these principles with the test techniques described in **Chapter 4**.

Testing shows the presence, not absence, of defects

Running a test through a software system can only show that one or more defects exist. Testing cannot show that the software is error-free. Consider whether the top 10 wanted criminals website was error-free. There were no functional defects, yet the website failed. In this case the problem was non-functional and the absence of defects was not adequate as a criterion for release of the website into operation.

In **Chapter 2** we will discuss retesting, when a previously failed test is rerun to show that under the same conditions the reported problem no longer exists. In this type of situation, testing can show that one particular problem no longer exists.

Although there may be other objectives, usually the main purpose of testing is to find defects. Therefore, tests should be designed to find as many defects as possible. However, everyone (especially business stakeholders) needs to know that testing cannot necessarily find all defects in the system under test. Even if this could be achieved, no one would **know** that it had been done.

Exhaustive testing is impossible

If testing finds problems, then surely you would expect more testing to find additional problems, until eventually we would have found them all. We discussed exhaustive testing earlier when looking at the smartphone mapping app and concluded that for large complex systems exhaustive testing is not possible. However, could it be possible to test small pieces of software exhaustively and only incorporate exhaustively tested code into large systems?

> Exhaustive testing: A test approach in which all possible data combinations are used. This includes implicit data combinations present in the state of the software/data at the start of testing.

Consider a small piece of software where one can enter a password, specified to contain up to three characters, with no consecutive repeating entries. Using only Western alphabetic capital letters and completing all three characters, there are $26 \times 26 \times 26$ input permutations (not all of which will be valid). However, with a standard keyboard there are not $26 \times 26 \times 26$ permutations, but a much higher number: $256 \times 256 \times 256$, or 2^{24}. Even then, the number of possibilities is higher. What happens if three characters are entered and the 'delete last character' right arrow key removes the last two? Are special key combinations accepted, or do they cause system actions (Ctrl + P, for example)? What about entering a character and waiting 20 minutes before entering the other two characters? It may be the same combination of keystrokes, but the circumstances are different. We can also include the situation where the 20-minute break occurs over the change-of-day interval. It is not possible to say whether there are any defects until all possible input combinations have been tried.

Even in this small example, there are many, many possible data combinations to attempt. The number of possible combinations using a smartphone might be significantly less, but it is still large enough to be impractical to use all of them.

Unless the application under test (AUT) has an extremely simple logical structure and limited input, it is not possible to test all possible combinations of data input and circumstances. Rather than attempting to test exhaustively, test techniques (see **Chapter 4**), test case prioritisation (see **Chapter 5**) and risk-based testing (see **Chapter 5**) should be used to focus test efforts.

Early testing saves time and money

When discussing why software fails, we briefly mentioned the idea of early testing. This principle is important because, as a proposed deployment date approaches, time pressure can increase dramatically. There is a real danger that testing will be squeezed, and this is bad news if the only testing we are doing is after all the development has been completed. The earlier the testing activity is started, the longer the elapsed time available is for testing and debugging. Testers do not have to wait until software is available to test.

Work products are created throughout the SDLC, and we talk about these different work products later in this chapter. As soon as these are ready, we can test them. In **Chapter 2** we will see that requirement documents are the basis for acceptance testing, so the creation of acceptance tests can begin as soon as requirement documents are available. As we create these tests, they will highlight the contents of the requirements. Are individual requirements testable? Can we find ambiguous or missing requirements?

Many problems in software systems can be traced back to missing or incorrect requirements. We will look at this in more detail when we discuss reviews in **Chapter 3**. The use of reviews can break the error–defect–failure cycle. In early testing we are trying to find errors and defects before they are passed to the next stage of the development process. Early testing techniques attempt to show that what is produced as a system specification, for example, accurately reflects that which is in the requirement documents. Ed Kit, in *Software Testing in the Real World*, discusses identifying and eliminating defects at the part of the SDLC in which they are introduced. If an error/defect is introduced in the coding activity, it is preferable to detect and correct it at this stage. If a problem is not corrected at the stage in which it is introduced, this leads to what Kit calls 'errors of migration'. The result is rework. We need to rework not just the part where the mistake was made, but each subsequent part where the error has been replicated. A defect found at acceptance testing where the original mistake was in the requirements will require several work products to be reworked, and subsequently to be retested.

Studies have been done on the cost impacts of errors at the different development stages. While it is difficult to put figures on the relative costs of finding defects at different levels in the SDLC, **Table 1.1** does concentrate the mind. This is known as the cost escalation model.

Table 1.1 Comparative cost to correct errors

Stage error is found	Comparative cost
Requirements	$1
Coding	$10
Program testing	$100
System testing	$1,000
User acceptance testing	$10,000
Live running	$100,000

What is undoubtedly true is that the graph of the relative cost of early and late identification/correction of defects rises very steeply (**Figure 1.5**). The earlier a problem (defect) is found, the less it costs to fix.

Figure 1.5 Effect of identification time on cost of errors

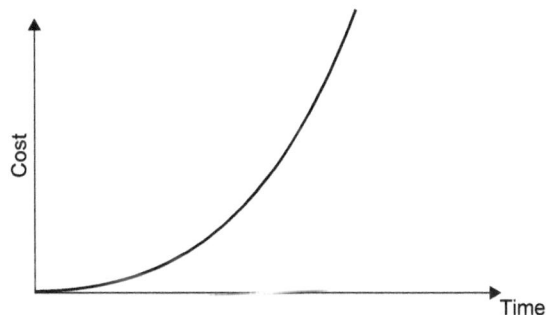

The objectives of various stages of testing can be different. For example, in the review processes we may focus on whether the documents are consistent and that no defects have been introduced when the documents were produced. Other stages of testing can have other objectives. The important point is that testing has defined objectives.

One of the drivers behind the push to Agile development methodologies is to enable testing to be incorporated throughout the software build process. This is nothing more than the 'early testing' principle.

Defects cluster together

Problems do occur in software. It is a fact. Once testing has identified (most of) the defects in a particular application, it is at first surprising that the spread of defects is not uniform. In a large application, it is often a small number of modules that exhibit the majority of the problems. This can be for a variety of reasons, including:

- system complexity;
- volatile code;
- the effects of change on change;
- development staff experience;
- development staff inexperience.

This is the application of the Pareto principle to software testing: approximately 80 per cent of the problems are found in about 20 per cent of the modules. It is useful if testing activities reflect this spread of defects, and target areas of the AUT where a high proportion of defects can be found. Prior to commencing testing, possible defect clusters can be predicted using the experience of testers and the factors listed above. This can be a useful input to risk-based testing (described in **Chapter 5**). However, it must be remembered that testing should not concentrate exclusively on these parts. There may be fewer defects in the remaining code, but testers still need to search diligently for them. If we have made predictions about (possible) defect clusters, what was expected and what is found may be different.

Tests wear out

Running the same set of tests continually will not continue to find new defects. Developers will soon know that the test team always tests the boundaries of conditions, for example, so they will learn to test these conditions themselves before the software is delivered. This does not make defects elsewhere in the code less likely, so continuing to use the same test set will result in decreasing the effectiveness of the tests. We can overcome this problem by changing existing tests and test data, using different test techniques and adding new tests. It is sometimes necessary to retire tests that are no longer appropriate.

For example, a small change to software could be specifically tested and an additional set of tests performed, aimed at showing that no additional problems have been introduced (this is known as regression testing). However, the software may fail in production because the regression tests are no longer relevant to the requirements of the system or the test objectives. Any regression test set needs to change to reflect business needs, and what are now seen as the most important risks. Regression testing is covered in more detail in **Chapter 2**.

Testing is context-dependent

Different testing is necessary in different circumstances. A website where information can merely be viewed will be tested in a different way to an ecommerce site where goods can be bought using credit/debit cards. We need to test an air traffic control system with more rigour than an application for calculating the length of a mortgage.

Risk can be a large factor in determining the type of testing needed. The higher the possibility of losses, the more we need to invest in testing the software before it is implemented. A fuller discussion of risk is given in **Chapter 5**.

For an ecommerce site, we should concentrate on security aspects. Is it possible to bypass the use of passwords? Can 'payment' be made with an invalid credit card by entering excessive data into the card number field? Security testing is an example of a specialist area, not appropriate for all applications. Such types of testing may require specialist staff and software tools. Test tools are covered in more detail in **Chapter 6**.

Absence-of-defects fallacy

Software with no known errors is not necessarily ready to be shipped. Does the AUT match up to the users' expectations of it? The fact that no defects are outstanding is not a good reason to ship the software. We may have undertaken **verification** of the software (the application does what it was meant to do, according to the stated requirements). However, are the requirements correct? In addition to verification, testing needs to check that the system is fit for purpose – we need to undertake **validation**. Can real users achieve what is required in an effective way?

Before dynamic testing has begun, there are no defects reported against the code delivered. Does this mean that software that has not been tested (but has no outstanding defects against it) can be shipped? We think not.

CHECK OF UNDERSTANDING

1. Why is having 'zero defects' an insufficient guide to software quality?
2. Give three reasons why defect clustering may exist.
3. Briefly justify the idea of early testing.

TEST ACTIVITIES, TESTWARE AND TEST ROLES

We previously determined that testing is a process. It would be easy to think that testing is thus always the same. However, this is not true.

Test activities and tasks

The most visible part of testing is running one or more tests: test execution. We also have to prepare for running tests, analyse the tests that have been run and see whether testing is complete. Both planning and analysing are necessary activities that enhance and amplify the benefits of the test execution itself. It is no good testing without deciding how, when and what to test. Planning is also required for the less formal test approaches, such as exploratory testing (covered in more detail in **Chapter 4**). We are describing a generalised test process – as was said earlier, this will not be the same on every project within an organisation, nor between different organisations. Not all of the activity groups that are described will be recognised as separate entities in some projects, or within certain organisations.

The following activity groups occur in the test process, which we will describe in more detail (**Figure 1.6**):

1. test planning;
2. test monitoring and control;
3. test analysis;
4. test design;
5. test implementation;
6. test execution;
7. test completion.

Figure 1.6 A generalised test process

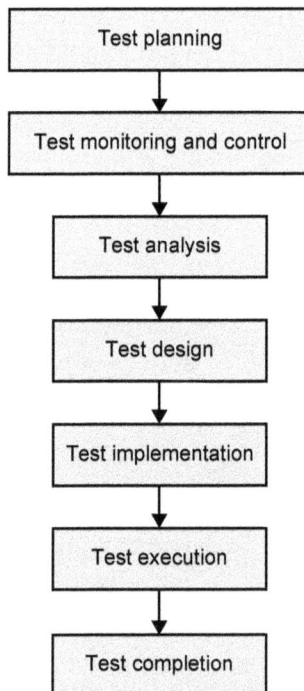

Although the main activity groups are in a broad sequence, they are not undertaken in a rigid way. The first two groups are overarching, and the other groups are both planned and monitored, and what has been undertaken may have a bearing upon other actions, some of which have been completed. An earlier activity group may need to be revisited. A defect found in test execution can sometimes be resolved by adding functionality that was originally not present (either missing or the new functionality is needed to make the other part correct). The new features themselves have to be tested, so even though implementation and execution are in progress, the 'earlier' activity groups of test analysis and test design have to be performed for the new features (**Figure 1.7**). The activity groups may overlap or be performed concurrently. It could be that test design is being undertaken for one test level at the same time as test execution is underway for another test level on the same project (test execution for system testing and test design for acceptance testing, for example). Please note, not all possible interactions between later and earlier groups have been included, and all of the latter five stages have interactions between test planning and test monitoring and control.

Figure 1.7 Iteration of activities

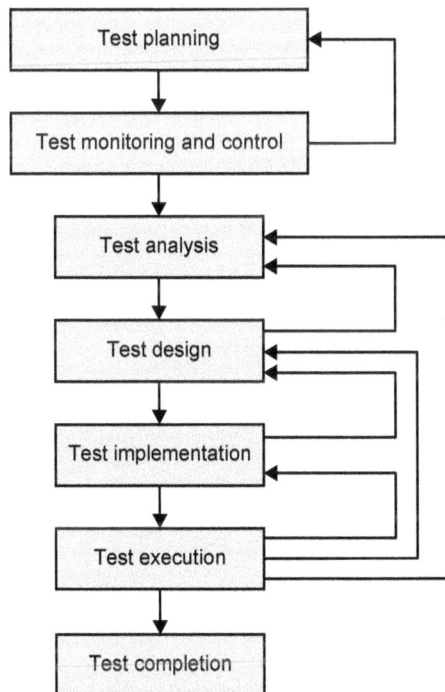

We sometimes need to do two or more of the main activities in parallel. Time pressure can mean that we begin test execution before all tests have been designed.

Test planning

Planning is determining what is going to be tested and how this will be achieved. It is where we draw a map for how activities will be done and who will do them. Test planning is also where we define the test completion criteria. Completion criteria are how we know when testing is finished. It is where the objectives of testing are set, and where the approach for meeting test objectives (within any context-based limitations) are set. A test plan is created and schedules are drawn up to enable any internal or external deadlines to be met. Plans and schedules may be amended, depending upon how other activities are progressing – amendments take place as a result of monitoring and control activities. The activities of test planning can be neatly summarised in the following six key questions:

Why?	*Test objectives*
Who?	*Resources (people)*
What?	*Scope (in and out)*
When?	*Schedule; entry/exit criteria*
Where?	*Environments and locations*
How?	*Test approach*

Planning is described in more detail in **Chapter 5**.

Test monitoring and control

Test monitoring and control go together. Monitoring is concerned with seeing if what has been achieved is what was expected to be done at this point in time, whereas control is taking any necessary action to meet the original or revised objectives as given in the test plan. Monitoring and control are supported by referring to the exit criteria or completion criteria for different stages of testing – this is sometimes referred to as the definition of 'done' in some project life cycles. Looking at exit criteria for test execution for a stage of testing may include:

- checking test results to see if the required test coverage has been achieved;
- determining the component or system quality by looking at test results and test logs;
- seeing if more tests are required (for example, if there are not enough tests to reach the level of risk coverage that is required).

Progress against the (original or revised) test plan is communicated to the necessary people (project sponsor, user stakeholders, the development team) in test progress reports. These will usually detail any actions being put in place to enable some milestones to be met earlier than would otherwise be the case or help to inform any decisions that stakeholders need to make. Test monitoring and control are covered more fully in **Chapter 5**.

Test analysis

Test analysis examines the test basis (which is usually written but not always confined to a single document) to identify what to test. Testable parts are identified and test conditions drawn up. Major activities in the test analysis activity group are as follows:

- Examine the test basis relevant for the testing to be carried out at this time:
  - requirements documents, functional requirements, business requirements, system requirements or other items (use cases, user stories) that detail both functional and non-functional component or system behaviour;
  - design or implementation detail (system or software architecture, entity-relationship diagrams, interface specifications) that define the component or system structure;
  - implementation information for the component or system, including code, database metadata and queries;
  - risk analysis reports, which may consider both functional and non-functional aspects, and the structure of the software to be tested.
- Look at the test basis, searching for defects of various types:
  - ambiguities;
  - things that have been missed out;
  - inconsistencies;
  - inaccuracies;
  - contradictions;
  - unnecessary or superfluous statements.
- Identify features and feature sets to be tested.
- Identify and prioritise test conditions for each feature or set of features (based on the test basis). Prioritisation is based on functional, non-functional and structural factors, other matters (both business and technical) and the levels of risk.
- Capture traceability in both directions ('bi-directional traceability') between the test basis and test conditions.

It can be appropriate and very helpful to use some of the test techniques that are detailed in **Chapter 4** in the test analysis activity. Test techniques (black box, white box and experience-based techniques) can assist to define more precise and accurate test conditions, and include important but less obvious test conditions.

Sometimes, test conditions produced as part of test analysis are used as test objectives in test charters. Test charters can be important in exploratory testing, discussed as part of experience-based test techniques in **Chapter 4**. In such situations, experience can be part of the test basis.

The test analysis activity can identify defects in the test basis. This can be especially important when there is no additional review process in place, or where the test process is closely aligned to the review process. Test design activity by its nature looks at the test basis (including requirements), so it is appropriate to see whether requirements are consistent, clearly expressed and not missing anything. The work of test analysis can also rightly ask whether formal requirements documents have included customer, user and other stakeholder needs. Some development methodologies involve creating test conditions and test cases prior to any coding taking place. Examples of such methodologies include behaviour-driven development (BDD) and acceptance test-driven development (ATDD). Both BDD and ATDD as development approaches are discussed in further detail in **Chapter 2**.

Test design

Test analysis answers the question 'What to test?', while test design answers the question 'How to test?' It is during the test design activity that test conditions are used to create test cases. This may also use test techniques, and the process of creating test cases can identify defects. Creating test cases involves the identification of coverage items, something that needs to be tested. If we have a clear boundary between different expected behaviours, we can apply boundary value analysis (BVA) in the creation of test cases (see **Chapter 4**). As we discussed earlier, a test case not only describes what to do to see if the test condition is met, but also what the expected result should be. Creating two or more test cases for a test condition can identify actions where the expected result is not clear – in the example we used earlier, are there different error messages if the username is incorrect and the password is incorrect? If the requirements, specification or other documents are not clear, there is potential for misunderstandings.

Activities in the test design activity group can be summarised as follows:

- using test conditions to create test cases, and prioritising both test cases and sets of test cases;
- identifying any test data to be used with the test cases;
- designing the test environment if necessary, and identifying any other items that are needed for testing to take place (infrastructure and any tools required);
- detailing bi-directional traceability between test basis, test conditions, test cases and test procedures – building upon the traceability that was detailed above under test analysis.

Test implementation

We saw that test design asks 'How to test?' Test implementation asks 'Do we have everything in place to run the tests?' It is the link between test design and running the test, or test execution. We have the tests ready; how can we run them? Test implementation is not always a separate activity; it can be combined with test design or, in exploratory testing and some other kinds of experience-based testing, test design and test implementation may take place and be documented as part of the running of tests. In exploratory testing, tests are designed, implemented and executed simultaneously.

As we get ready for test execution, the test implementation main activities are as follows:

- Creating and prioritising test procedures and possibly creating automated test scripts.

- Creating test suites from the test procedures and from automated test scripts if there are any.

- Sequencing test suites in a test execution schedule, so that it makes efficient use of resources. For example, a test execution schedule could **create** a customer order, **amend** that order and then **delete** it – all done to run efficiently.

- Building the test environment with anything extra in place (test harnesses, simulators, dummy third-party interfaces, service virtualisation and any other infrastructure items). The test environment then has to be checked to ensure that all is ready to start testing.

- Preparing test data and checking that it has been loaded into the test environment.

- Checking and updating the bi-directional traceability between the test basis, test conditions, test cases, test procedures and now to include test suites.

Test execution

Test execution involves running tests, and is where (what are often seen as) the most visible test activities are undertaken. Test suites are run as detailed in the test execution schedule (both of which were created as part of test implementation). Execution includes the following activities:

- Recording the identification and version of what is being tested: test items or test object, test tool(s) and any testware that are in use. This information is important if any defects are to be raised.

- Running tests, either manually or using test execution tools.

- Comparing actual and expected results (this may be done by any test execution tool(s) being used).

- Looking at instances where the actual result and the expected result differ to determine the possible cause. This may be a defect in the software under test, but could also be that the expected results were incorrect or the test data was not quite correct.

- Reporting defects based on the failures that were found in the testing.

- Recording the result of each test (pass, fail, blocked).

- Repeating tests where the software has been corrected, the test data has been changed or as part of regression testing.

- Confirming or amending the bi-directional traceability between the test basis, the test conditions, test cases, test procedures and expected results.

Test completion

Testing at this stage has finished, and test completion activities collect the data so that lessons can be learned, testware reused in future projects and so on. These activities may occur at the time that the software system is released, but could also be at other significant project milestones – the project is completed (or even cancelled), a test level is completed or an Agile project iteration is finished (where test completion activities may be part of the iteration retrospective meeting). The key here is to make sure that information is not lost (including the experiences of those involved). Activities in the test completion group are summarised below:

- Checking that defect reports are all closed as necessary. This may result in the raising of change requests or creating product backlog items for any that remain unresolved when test execution activities have ended.

- Creating a test completion report to communicate the results of the testing activities. This is usually for the benefit of the project stakeholders.

- Closing down and saving ('archiving') the test environment, any test data, the test infrastructure and other testware. These may need to be reused at a later date.

- Handing over the testware to those who will maintain the software in the future or any other projects or other stakeholders to whom it could be beneficial.

- Analysing lessons learned from testing activities that have been completed, which will hopefully drive changes in future iterations, releases and projects.

- Using the information that has been gathered to make the test process better – to improve test process maturity.

Test process in context

Variations between organisations, and indeed projects within the same organisation, can have an influence on how testing is carried out, and on the specific test process that is used. Specific matters, or contextual factors, that affect the test process can be many and varied, so please do not assume that the factors listed in the syllabus are the **only** factors. Many of these are covered in more detail in later chapters of this book. We give those in the syllabus here; remember, you could be examined on these:

- stakeholders (needs, expectations, requirements, willingness to cooperate, etc.);

- team members (skills, knowledge, level of experience, availability, training needs, etc.);

- business domain (criticality of the test object, identified risks, market needs, specific legal regulations, etc.);

- technical factors (type of software, product architecture, technology used, etc.);

- project constraints (scope, time, budget, resources, etc.);

- organisational factors (organisational structure, existing policies, practices used, etc.);

- SDLC (engineering practices, development methods, etc.);

- tools (availability, usability, compliance, etc.).

These items (and others not listed in the syllabus that you may be aware of) can have a very significant impact on the way that testing is carried out and on multiple testing activities, including the test techniques used, the level of any test automation, test reporting, the level of documentation produced in the testing effort and the test strategy.

Agile methodologies and a generalised test process

Thus far we have concentrated on 'traditional' methodologies when discussing the test process. We now focus on Agile methodologies.

The use of Agile methodologies and the relationship with a test process is not a syllabus topic for the Foundation Certificate, but in the following discussion our understanding of both a generalised test process and Agile methodologies will grow.

We use the term 'Agile methodologies' because there is not a single variant or 'flavour' of Agile. The distinctions between these are unimportant here, but the principles are important. Agile SDLCs aim to have frequent deliveries of software, where each iteration or sprint will build on any previously made available. The aim is to have software that could be implemented into the production environment at the end of each sprint – although a delivery into production might not happen as frequently as this.

From this brief introduction to Agile, it follows that an Agile project has a beginning and an end, but the processes between these stages can be repeated many times. Agile projects can successfully progress over the course of many months, or even years, with a continuous stream of (same length) iterations producing production-quality software, typically every two, three or four weeks. In terms of the test process, there is a part of the test planning stage that takes place at the start of the project, and the test completion activities take place at the end of the project. However, some test planning and all of the middle five stages of the test process we have described above are present in every sprint.

Of the test planning activities taking place at the beginning of the project, typically included are resourcing activities, some outline planning on the length of the project and an initial attempt at ordering the features to be implemented (although this could change significantly as the project progresses, sprint by sprint). Infrastructure planning, together with the identification and provision of any specific testing tools, is usually undertaken at this stage, together with a clear understanding within the whole team of what is 'done' (i.e. a definition of 'done').

At the start of each sprint, planning activity takes place to determine items to include in the sprint. This is a whole team activity based on reaching a consensus of how long each potential deliverable will take. As sprint follows sprint, so the accuracy of estimation increases. The sprint is a fixed length, so throughout the duration of the development, items could be added or removed to ensure that, at the conclusion, there is tested code 'on the shelf' and available for implementation as required.

Other activities of the generalised test process we described are undertaken in each sprint. A daily stand-up meeting should result in a short feedback loop to enable any adjustments to take place and items that prevent progress to be resolved. The whole time for development and testing is limited, so the preparation for testing and the testing itself have to be undertaken in parallel. Towards the end of the sprint it is possible that most or all of the team are involved in testing, with a focus on delivering all that was agreed at the sprint planning meeting.

Automated testing tools are frequently used in Agile projects. Tests can be very detailed, and a tester on such a project can provide very useful input to developers in defining tests to be used to identify specific conditions. The earlier in a sprint that this is done, the more advantages can be gained within that sprint.

The proposed deliverables are sometimes used in conjunction with the definition of 'done' to enable a burn-down chart to be drawn up. This enables a track to be kept of progress for all to see – in itself usually a motivating factor. As sprint follows sprint, the subject of regression testing previously delivered, working software becomes more important. Part of the test analysis and test design will involve selecting regression tests that are appropriate for the current sprint. Tests that previously worked might now (correctly) not pass because the new sprint has changed the intention of previously delivered software. The conclusion of the sprint is often a sprint review meeting, which can include a demonstration to user representatives and the project sponsor.

For development using Agile methodologies, the final stage of our test process – 'test completion activities' – is scheduled after the end of the last sprint. This should not be done earlier because testing collateral that was used in the last-but-three sprint might no longer be appropriate at the conclusion of the final sprint. As we discussed earlier, even the regression tests might have changed, or a more suitable set of regression tests identified.

Further information about Agile methodologies is given in **Chapter 2**.

Test work products

Each of the activity groups we have discussed has one or more work products that are typically produced as part of that set of activities. However, just as there are major variations in the way particular organisations implement the test process, so there can be variations in both the work products that are produced and in some cases even the names of those work products. This section follows the test process that we outlined above, and work products are given for each of the seven activity groups that we described in a generalised test process.

Many of the work products that are described here can be captured and managed using test management and defect management tools (these two tool types are both described in more detail in **Chapter 6**). Work products used in test process activity groups can often be easily attributed to the appropriate activity group or when considering the preceding or following activity group, so even though 'test work products' can be a topic in the examination, this does not mean that you have to learn the lists that are given. An example of an examination-type question relating to this area of study is given at the end of this chapter.

There are some work products that are typically created in two or more test process activity groups. Examples of this are defect reports (most usually in test analysis, test design, test implementation and test execution) and test reports (test progress reports and test summary reports in test monitoring and control, test completion reports in both test monitoring and control, and test completion).

Test planning work products

Test planning activities usually produce plans and schedules. There can be several test plans for a project: component testing test plan, integration testing test plan and so on. Test plans contain information about the test basis and the exit criteria (or definition of 'done'), so that we know in advance when we can say that testing is complete. Other work products will be related to the test basis by traceability information. A risk register may be produced as a work product from test planning; this is a list of risks, together with their likelihood, the impact of the risks and possible ways of lessening these (risk mitigation actions). Risk is covered in **Chapter 5**.

Test monitoring and control work products

Work products produced from the test monitoring and control activities include periodic test progress reports. Not all reports are for the same target audience, so any reports need to be audience-specific in the level of detail. Reports for senior stakeholders may just give the number of tests that have passed, have failed and cannot yet be run, for example, whereas reports for the wider development team may include subsystem-specific information about the tests that have passed and failed.

Test monitoring and control work products also need to highlight project management concerns, including task completion, the use of resources and the amount of effort that has been expended. Where necessary, there may be choices highlighted with possible actions that can be taken to still achieve the original timescales, even though at the present time a project is behind schedule. Monitoring and control are part of a feedback loop; monitoring highlights matters and control aims to address these items. Information found in test monitoring can be used in product risk analysis and mechanisms to address risks, as detailed in **Chapter 5**.

Test analysis work products

The following are work products that the test analysis group of activities is expected to produce:

- Defined and prioritised test conditions, ideally each with bi-directional traceability to the specific part(s) of the test basis that is covered.
- Acceptance criteria is an example used with test conditions in Agile life cycles (see **Chapter 4**).
- Test analysis may result in the discovery and reporting of defects in the test basis.

Test design work products

The key work products that come from test design activities are test cases and groups of test cases that relate to the test conditions created by test analysis. It is extremely useful if these are bi-directionally traceable to the test conditions they relate to. Test cases can be high-level test cases (without specific values or concrete values to be used for input values) or low-level test cases (with detailed input values and **specific expected results**). High-level test cases can be reused in different test cycles, although they are not as easy to use for individuals who are not familiar with the software being tested.

At the test design stage, test charters are created for exploratory testing sessions. Other work products that are produced or amended as part of test design include amendments to the test conditions created in test analysis, the design and identification of any test data that is needed, the definition and design of the test environment and the identification of any infrastructure or tools that may be required. However, although these are done, the amount that is documented can vary.

Test implementation work products

Test implementation activities have some work products that are more easily identifiable than others. The first three below are very recognisable, but the others can also be created at this stage:

- Test procedures and the sequencing of these.
- Test suites, comprising two or more test procedures.
- A test execution schedule.
- In some cases, test implementation activities create work products using or used by tools:
  - service virtualisation;
  - automated test scripts.
- Creation and verification of the test data.
- Creation and verification of the test environment.
- Further refinement of the test conditions that were produced as part of test analysis.

When we have specific test data, this can be used to turn high-level test cases into low-level test cases, and to these are added expected results. In some instances, expected results can be derived from the test data using a test oracle – you feed the data in and get the expected results as the answer.

Once test implementation is complete it can be possible to see the level of coverage by written test cases for parts or all of the test basis. This is because, at each stage, we have built-in bi-directional traceability. It may be possible to see how many test procedures have been written that cover a specific requirement or area of functionality.

Test execution work products

The three key work products created by test execution activities are given below:

- The status of individual tests (passed, failed, skipped, ready to run, blocked, etc.).
- Defect reports as a result of test execution.

- Test logs, being a chronological record of the relevant details about the execution of tests. These can be very useful (and important) if and when tests need to be repeated, be that as a result of defect correction, because the environment has changed or other reasons, which would include regression testing.

When testing is complete, if there is bi-directional traceability between the test basis, test conditions, test cases, test procedures and test suites, it is possible to work backwards and say which requirements have failed tests against them, where the impact of defects touches the business areas and so on. This will enable checking that the predetermined coverage level has been met, or not, and help to report in ways that business stakeholders understand.

Test completion work products

When testing is complete, the following can be created from the test completion activities:

- test completion reports;
- a list of improvements for future work (the next iteration or next project);
- change requests or product backlog items;
- finalised testware (and sometimes the test environment) for future usage.

Traceability between the test basis and test work products

Throughout the descriptions of the test process groups of activities and the work products that are created by these activity groups, there has been a sub-story of bi-directional traceability. Primarily this enables the evaluation of test coverage – the extent to which the test objectives have been achieved – but also how does this one failed test reflect back to requirements or business goals? There are several great advantages to having full traceability:

- assessing the impact of changes (if one requirement changes, how many test conditions, test cases, test procedures and test suites may be affected);
- making testing auditable;
- enabling IT governance criteria to be met;
- improving the understandability of test progress reports and test summary reports to reflect passed, failed and blocked tests back to requirements or other aspects of the test basis;
- relating testing to stakeholders in terms that they can understand (it is not 'three failed tests' but 'three tests failed, which means that creating a customer is not possible');
- enabling the assessment of product quality, process capability and project progress against the goals of the business.

Roles in testing

There are many titles that could be used to describe the activities of those involved in the test process (among these could be test architect, test coordinator, test guru, test analyst, etc.). The syllabus concentrates on two of these: a test management role and a tester role. These may not be precisely those used in any specific organisation you are aware of, but they are convenient divisions to use, partitioning the areas of responsibility.

The test management role directs and has overall control of the testing activity, and will lead testing in terms of decisions and direction. This role is primarily concentrated on the activities of test planning, test monitoring and control and test completion. There are no set rules for how test management is carried out. We discussed earlier the contextual matters that will have an impact on testing. It is in this role that these impacts are felt. Not all things related to the testing effort will be determined by the test management role on a particular project. There may be company-wide standards on test automation, or the use of test techniques, covered by an organisational test policy (outside the scope of this syllabus).

The tester role will concentrate on test analysis, test design, test implementation and test execution. The role is responsible for the engineering aspects of testing.

Within these two broad categories there can be many job titles that you are familiar with. It is also possible for different people to take on these roles at different times. The test manager role can be performed by a variety of people (e.g. a team leader, a test manager, a development manager). Some tasks from the test management role can be handled by 'the team' when there is an Agile development methodology.

CHECK OF UNDERSTANDING

1. What are the activity groups in the generalised test process described (in the correct sequence)?
2. Give advantages of maintaining traceability between the test basis and test work products (including test cases).
3. When should the expected outcome of a test be defined?
4. Give three work products that are created during the test execution group of activities.
5. Is it the test management role or the tester role that is mainly focused on test design?

ESSENTIAL SKILLS AND GOOD PRACTICES IN TESTING

A variety of different people may be involved in the total testing effort, and they may be drawn from a broad set of backgrounds. Some will be developers, some professional testers and some specialists, such as those with performance testing skills, while others may be users drafted in to assist with acceptance testing. Whoever is involved in testing needs at least some understanding of the skills and techniques of testing to make an effective contribution to the overall testing effort. In this section we will look at the generic skills required for testing, the idea of a whole team approach for testing and testing independence.

Generic skills

Testers are part of a team, the team that is tasked with providing a product that achieves (or attempts to achieve) what users require. As such, they need to play a full part in that team, concentrating on the end goal – working software to deliver business benefit in a timely and cost-effective way. The whole team usually comprises not only testers, but business users, technical architects, developers and potentially many more individuals. Some of the skills are really required by all team members, some only by testers. What is it that ideally testers bring as their contribution to the team? There are six skills that are of particular importance to testers.

Testing knowledge
How to test software, with the need to plan what will be tested, when and how. In-depth knowledge of test techniques can improve the quality of testing, and perhaps find more defects with fewer test cases. We have already started to look at some of the testing knowledge required – more will follow in later chapters.

Attention to detail and working in a methodical way
Testers are usually detail-oriented individuals who notice when something is not as it should be. They are looking for things that should not happen. This requires concentration on the little things, for example where a calculation is adrift by a small but measurable amount.

Good communication skills, being an active listener and a team player
It is important to consider the recipients when conveying to developers details of a new defect found. A tester may have to overcome what is termed 'confirmation bias' – where an author cannot or will not see a problem in their completed work. It needs both tact and good communication skills. The latter is a skill that all require, and for many it must be learned and constantly worked upon. Testers can sometimes be the bearers of bad news ('the software does not work'). Good communication skills are needed to convey defects in a constructive way. Defect management is covered in more detail in **Chapter 5**. Testers need to be diplomats at times, highlighting a potential problem without pointing an accusing finger at those responsible for the creation of the item(s) being tested. It is always good to consider that the tester may have got it wrong. Perhaps the test performed correctly, and the reason the tester pointed out a 'problem' was that their own understanding was less than perfect.

Analytical thinking and a creative approach
Defects can be difficult to find, so there needs to be a detailed search for them, in both likely and unlikely locations. This can sometimes be called 'professional curiosity' and can manifest itself in asking 'I wonder what would happen if ...' This particular aspect is discussed in more detail in **Chapter 4** when describing exploratory testing.

Technical knowledge
This involves, for example, using test tools to increase the efficiency of testing, or knowing how to navigate the underlying database.

Domain knowledge

What the business is involved in, and how the software will help the business to achieve their goals. What are the most important transactions in terms of business value? Are these the most frequent? Who will use the software? Is this just the (internal) policy clerks, or will the software be available on the internet for anyone to use?

Whole team approach

Testers need to work effectively with others. The aim of everyone involved in the development process (including testers) is to build a successful product. Therefore, everyone needs to work together on the common goals. The whole team approach is very popular in Agile methodologies such as Scrum, and stems from its use in the extreme programming methodology. In this approach we see fewer people continuing in their specialised way of working; instead each task is undertaken by whoever is best placed at that time – the person testing now may not be a 'tester', but they are the most suitable person for the task in question. A key point is that **everyone is responsible for quality**. In a whole team approach, testers can educate others in the team about testing and transfer some of their testing knowledge.

The team share the same workspace, which aids communication and interaction. Testers work together with others, including the business, with a subsidiary aim to transfer testing knowledge to the rest of the team and influence the development of the product. This includes working with business stakeholders in creating suitable acceptance tests and agreeing with developers on the test strategy and deciding on test automation approaches.

The whole team approach is not always appropriate. In safety-critical software, tester independence may be required, with those involved in testing needing to be separate from those creating the work product.

Independence of testing

Testing can be more effective if it is not undertaken by the individual(s) who wrote the code, for the simple reason that the creator of anything (whether it is software or a work of art) has a special relationship with the created object. The nature of that relationship is such that flaws in the created object are sometimes rendered invisible to the creator. For that reason, it is important that someone other than the creator should test the object. Of course, we want the developer who builds a component or system to debug it, and even to attempt to test it, but we accept that testing done by that individual cannot be assumed to be complete. Developers can test their own code, but it requires a mindset change, from that of a developer (to prove it works) to that of a tester (trying to show that it does not work).

The idea of tester independence relates to how near testers are to those who create the work products that are being tested. This does not necessarily mean geographical location, although as a rule, where testers are less independent there tends to be nearness of physical location as well – but this is not always the case. However, geographical co-location does **not** always mean less testing independence. In Agile development projects the whole team can be positioned together.

On an ascending scale of independence, a software work product may be tested by:

- the author;
- a peer from the same team as the author;
- separate testers from the same project;
- a separate test team within the same company ('independent test team');
- testers from outside the organisation.

These are just examples of different levels of testing independence; there are not only five levels of independence. Some workplaces will have more levels, and some will have fewer, and distinctions are not always as rigid as given here.

On many software projects there are multiple levels of independence. Testing can happen at different stages in the development life cycle, and those involved at these times are not necessarily the same people.

Greater tester independence can have advantages and disadvantages. These should be kept in mind when determining whether more independence is required in particular situations. Advantages of greater levels of independence include:

- Different kinds of defects and potential failures can be found for many reasons:
  - The testing experience of those involved.
  - A noteworthy defect found previously can ensure that particular types of test are always performed.
  - Testers removed from the development environment will not necessarily have the same assumptions or can more easily challenge assumptions.
- Independence can overcome people missing defects because of the 'confirmation bias' – you see what you want to see.
- Testers outside the development team have little or no emotional involvement in the items being tested.
- Those external to the development team are not part of the organisational internal politics surrounding that team and the people in it.

There are also disadvantages to greater levels of independence:

- Being isolated from the development team, which can make face-to-face communications more difficult.
- There can be a lack of collaboration, communication problems and sometimes an adversarial relationship between the teams.
- Testing can be perceived as a bottleneck or blamed for delays in the release of software.
- If external teams (within or outside the organisation) perform the testing, developers can lose the sense of responsibility for the quality of the work produced: 'Oh, its alright. The testers will find any problems with the software.'

- An external test team may have to spend considerable time getting to understand the software and how it should perform. The test team must get to know the product, and should prioritise tests for situations that are likely to happen, rather than concentrate on scenarios that are at best 'unlikely'.

CHECK OF UNDERSTANDING

1. Give three key skills that each tester should have. Explain how each one has an impact on the performance of a tester.

2. Contrast the advantages and disadvantages of developers testing their own code.

3. How can a tester give bad news in a way that is not confrontational?

CODE OF ETHICS

This section is **not** examinable but is retained (being in earlier versions of the syllabus but not the current one) as it is a useful topic for the tester and aspiring tester to be aware of. The content that follows is based on the Association for Computing Machinery (ACM) Code of Ethics and on the IEEE Code of Ethics for Engineers.

We will look at how testers should behave as professionals in the workplace, a code of ethics, before we move onto the more detailed coverage of topics in the following chapters. Testers can have access to confidential and privileged information, and they are to treat any information with care and attention, and act responsibly when dealing with the owner(s) of this information, employers and the wider public interest. Of course, anyone can test software, so the declaration of this code of ethics applies to those who have achieved software testing certification. The code of ethics applies to the following areas:

- Public – certified software testers shall consider the wider public interest in their actions.

- Client and employer – certified software testers shall act in the best interests of their client and employer (being consistent with the wider public interest).

- Product – certified software testers shall ensure that the deliverables they provide (for any products and systems they work on) meet the highest professional standards possible.

- Judgement – certified software testers shall maintain integrity and independence in their professional judgement.

- Management – certified software test managers and leaders shall subscribe to and promote an ethical approach to the management of software testing.

- Profession – certified software testers shall advance the integrity and reputation of the profession consistent with the public interest.

- Colleagues – certified software testers shall be fair to, and supportive of, their colleagues and promote cooperation with software developers.

- Self – certified software testers shall participate in lifelong learning regarding the practice of their profession and shall promote an ethical approach to the practice of the profession.

The code of ethics is far-reaching in its aims, and a quick review of the eight points reveals interaction with specific areas of the syllabus. The implementation of this code of ethics is expanded on in all chapters of this book, and perhaps is the reason for the whole book itself.

SUMMARY

In this chapter we have looked at key ideas that are used in testing and introduced some terminology. We examined some of the types of software problems that can occur, and why the blanket explanation of 'insufficient testing' is unhelpful. The problems encountered then led us through some questions about the nature of testing, why errors and mistakes are made and how these can be identified and eliminated. Individual examples enabled us to look at what testing can achieve, and the view that testing does not improve software quality but provides information about that quality.

We have examined both general testing principles and a standard template for testing: a generalised test process. These are useful and can be effective in identifying the types of problems we considered at the start of the chapter. The chapter finished by looking at the skills required by testers and the idea of tester independence, and finally covered the code of ethics for testers (non-examinable).

This chapter is an introduction to testing, and to themes that are developed later in the book. It is a chapter in its own right, but also points to information that will come later. A rereading of this chapter when you have worked through the rest of the book will place all the main topics into context.

Example examination questions with answers

E1. K1 question
What is a test condition?

    A.   A set of test data written to exercise one or more logic paths through the software under test.

    B.   A testable aspect of a component or system identified as a basis for testing.

    C.   The body of knowledge used as the foundation for test analysis and design.

    D.   The predetermined goals that will enable a decision to be made about whether testing is complete.

E2. K2 question
Which of the following definition pairs for testing and debugging is correct?

A. Debugging can show failures in the software; testing is investigating the causes of any failures and performing corrections.

B. Debugging is investigating the causes of software failures and undertaking corrective action; testing attempts to uncover problems by executing the software.

C. Debugging is always undertaken by developers; testing can be performed by development or testing personnel.

D. Debugging checks that software fixes have been resolved; testing is looking for any unintended consequences of a software fix.

E3. K2 question
Which of the following skills are appropriate for each tester?

i. Has knowledge of the business and how the software will be used to accomplish business goals.

ii. From a development background so can read code.

iii. Good communication skills (both spoken and written).

iv. Has a methodical and careful way of working so that all items are covered.

A. i, ii and iv.
B. ii, iii and iv.
C. i, iii and iv.
D. i, ii and iii.

E4. K2 question
Which of the following illustrates one of the testing principles?

A. No unresolved defects does not mean the software will be successful.

B. The more you test, the more defects will be found.

C. All software can be tested in the same way using the same test techniques.

D. Defects are usually found evenly distributed throughout the software under test.

E5. K2 question
Which of the following activities are part of the test implementation activity group, and which part of test execution?

    i.   Developing and prioritising test procedures, creating automated test scripts.

    ii.   Comparing actual and expected results.

    iii. Verifying and updating bi-directional traceability between the test basis, test conditions, test cases, test procedures and test results.

    iv. Preparing test data and ensuring it is properly loaded in the test environment.

    v.   Verifying and updating bi-directional traceability between the test basis, test conditions, test cases, test procedures and test suites.

    A.   i, ii and iii are part of test implementation; iv and v are part of test execution.

    B.   i, iii and v are part of test implementation; ii and iv are part of test execution.

    C.   i, ii and iv are part of test implementation; iii and v are part of test execution.

    D.   i, iv and v are part of test implementation; ii and iii are part of test execution.

E6. K1 question
Which of the following describes how testing relates to quality?

    A.   Testing is a form of quality control, a product-oriented corrective approach.

    B.   Testing is a form of quality control, a process-oriented preventive approach.

    C.   Testing is a form of quality assurance, a process-oriented preventive approach.

    D.   Testing is a form of quality assurance, a product-oriented corrective approach.

E7. K2 question
Which of the following is a recognised reason for testing to be carried out?

    A.   Using testers in the review of requirements will verify that the software is fit for purpose.

    B.   Evidence suggests that 25–45 per cent of the project costs should be used in the test process.

    C.   The detection and removal of defects increases the likelihood that the software meets stakeholder needs.

    D.   Testing ensures that there are no residual defects in the software under test.

E8. K2 question
Problems persist with the online customer interface for a utilities company after the introduction of smart meters in customer homes. Which *one* of the following is a root cause of a defect rather than an effect of a defect?

A. Consumption usage for some days is shown as zero, while that for other days is abnormally high.

B. Some chosen menu options intermittently display the 'system busy' icon, but never the correct information at the time it is selected.

C. Customer consumption data is only displayed for up to three days ago, with year-to-date and month-to-date excluding the last three days.

D. Systems architects did not anticipate the amount of web traffic the introduction of smart meters would generate.

E9. K2 question
Which *two* of the following work products are created during the test implementation activity?

i. Documentation about which test item(s), test object(s), test tools and testware were involved in the testing.

ii. Test execution schedule.

iii. Test cases.

iv. Documentation about the **status** (e.g. 'pass', 'fail', 'not run', etc.) of individual test cases or procedures.

v. Test procedures and their sequencing.

A. iii and v.

B. ii and v.

C. i and iii.

D. ii and iv.

Answers to the self-assessment questions in the chapter

SA1. The correct answer is C.

SA2. The correct answer is B.

SA3. The correct answer is D.

Answers to example examination questions

E1. The correct answer is B.

      A.   is the definition of a test case.
      C.   is the description of a test basis.
      D.   is a broad description of test exit criteria.

E2. The correct answer is B.

      A.   the two parts are 'switched'; what is described as 'debugging' is in fact 'testing', and vice versa.
      C.   is not true. The syllabus states that in some life cycles testers may be involved in debugging.
      D.   indicates that debugging is involved in retesting software after fixes. This is a description of 'retesting', a testing activity. The description given to 'testing' is that of regression testing; testing involves more than this.

E3. The correct answer is C.

Each of the options gives **three** of the **four** choices, leaving **one** incorrect choice. This question is asking a negative question in a positive way, and examinations for this qualification have used this type of question in the past. Choice ii is the only one of those given that is not included in the syllabus as 'generic skills required for testing'. Option C is the answer that does not include this choice. Note the question asks for each tester. Having someone in the test team that can read code (that is, from a development background) has advantages for the working of the team. To have everyone from the same background may be unwise.

E4. The correct answer is A.

      B.   this option has some truth to it, but it is not one of the testing principles. However, if there are no defects in the code, no amount of testing will find defects.
      C.   is not true. It is in direct contradiction to the 'testing is context-dependent' principle.
      D.   is again not true. This is usually found **not** to be the case, being the opposite of the 'defect clustering' principle.

E5. The correct answer is D.

Two of the choices are very similar, choices iii and v. The differences here is test results (choice iii) and test suites (choice v). This points to choice iii being part of test execution and choice v being part of test implementation. Option D is the only one that has these assigned in this way.

E6. The correct answer is A.

Quality assurance is a process-oriented activity, whereas quality control is a product-oriented one, as given in the syllabus. This rules out options B and D. One of the prime focuses of testing (not the only focus) is to find errors and correct them – a corrective approach. This points to option A as the correct answer.

E7. The correct answer is C.

Reviewing **requirements** can never verify that **software** is fit for purpose (option A). Just because evidence **may** suggest that a proportion of project cost should be spent on testing is not a reason to perform testing. This rules out option B. Option D states that testing can find all defects, which means that we can show that there are no remaining defects – contrary to one of the testing principles about testing only finding defects; it cannot show that are **no defects**. This leaves option C, which is the correct answer.

E8. The correct answer is D.

A root cause is **why** something has happened, as opposed to **what** has happened. Options A, B and C describe unusual events (which may or may not be defects but are certainly irritating for customers). Option B could be a system overload problem – this may be as a result of a higher than expected level of web traffic. This root cause is described in option D, the correct answer.

E9. The correct answer is B.

We will consider each of the work products in turn:

i. Documentation about which test item(s), test object(s), test tools and testware were involved in the testing – **test execution**.
ii. Test execution schedule – **test implementation**.
iii. Test cases – **test execution**.
iv. Documentation about the **status** (e.g. 'pass', 'fail', 'not run', etc.) of individual test cases or procedures – **test execution**.
v. Test procedures and their sequencing – **test implementation**.

Option B gives the correct choices.

2 LIFE CYCLES

Angelina Samaroo

INTRODUCTION

In the previous chapter we looked at testing as a concept – what it is and why we should do it. In this chapter we will look at testing as part of the overall software development process. Clearly, testing does not take place in isolation; there must be a product first.

We will refer to work products and products. A work product is an intermediate deliverable required to create the final product. Work products can be documentation or code. The code and associated documentation will become the product when the system is declared ready for release. In software development, work products are generally created in a series of defined stages, from capturing a customer requirement, to creating the system, to delivering the system. These stages are usually shown as steps within a software development life cycle (SDLC).

In this chapter we will look at:

- testing within software development models;
- test levels and test types;
- maintenance testing.

Learning objectives

The learning objectives for this chapter are listed below. You can confirm that you have achieved these by using the self-assessment questions immediately following the learning objectives, the 'Check of understanding' boxes distributed throughout the text and the example examination questions provided at the end of the chapter. The chapter summary will remind you of the key ideas.

Each learning objective is allocated a K number to represent the level of understanding required; see the **Introduction** (pp. 2–3) for an explanation of K numbers.

Software development life cycle models

- FL-2.1.1 (K2) Explain the impact of the chosen SDLC on testing.
- FL-2.1.2 (K1) Recall good testing practices that apply to all SDLCs.
- FL-2.1.3 (K1) Recall the examples of test-first approaches to development.

- FL-2.1.4 (K2) Summarise how DevOps might have an impact on testing.
- FL-2.1.5 (K2) Explain the shift-left approach.
- FL-2.1.6 (K2) Explain how retrospectives can be used as a mechanism for process improvement.

Test levels and test types

- FL-2.2.1 (K2) Distinguish the different test levels.
- FL-2.2.2 (K2) Distinguish the different test types.
- FL-2.2.3 (K2) Distinguish confirmation testing from regression testing.

Maintenance testing

- FL-2.3.1 (K2) Summarise maintenance testing and its triggers.

Self-assessment questions

The following questions have been designed to enable you to check your current level of understanding for the topics in this chapter. The answers are at the end of the chapter.

Question SA1 (K2)
Which of the following is true of the V model?

    A. Coding starts as soon as each function in a system has been defined.
    B. The test activities occur after all development activities have been completed.
    C. It enables the production of a working version of the system as early as possible.
    D. It enables test planning to start as early as possible.

Question SA2 (K2)
Which of the following is true of white box testing?

    A. It is carried out only by developers.
    B. It can be used to test data flow structures.
    C. It is used only at unit and integration test levels.
    D. Coverage achieved using white box test techniques is not measurable.

Question SA3 (K1)
Which of the following is a test object for integration testing?

    A. A subsystem.
    B. An epic.
    C. A risk analysis report.
    D. A sequence diagram.

TESTING WITHIN SOFTWARE DEVELOPMENT MODELS

A development life cycle for a software product involves capturing the initial requirements from the customer, expanding on these to provide the detail required for code production, writing the code and testing the product, ready for release.

A simple development model is shown in **Figure 2.1**. This is known traditionally as the waterfall model.

Figure 2.1 Waterfall model

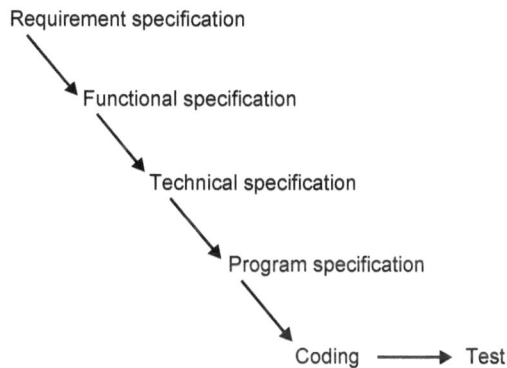

The waterfall model in **Figure 2.1** shows the steps in sequence, where the customer requirements are progressively refined to the point where coding can take place. This type of model is often referred to as a linear or sequential model. Each work product or activity is completed before moving on to the next.

In the waterfall model, testing is carried out once the code has been fully developed. Once this is completed, a decision can be made on whether the product can be released into the live environment.

This model for development shows how a fully tested product can be created, but it has a significant drawback: What happens if the product fails the tests? Let us look at a simple case study.

CASE STUDY: DEVELOPMENT PROCESS

Let us consider the manufacture of a smartphone. Smartphones have become an essential part of daily life for many. They must be robust enough to withstand the rigours of being thrown into bags or on floors and must be able to respond quickly to commands.

Smartphones have touchscreens. This means that the apps on the phone must be accessible via a tap on the screen. This is done via a touchscreen driver. The driver is a piece of software that sits between the screen (hardware) and the apps (software), allowing the app to be accessed from a tap on an icon on the screen.

If a waterfall model were to be used to manufacture and ship a touchscreen phone, then all functionality would be tested at the very end, just prior to shipping.

If it is found that the phone can be dropped from a reasonable height without breaking, but that the touchscreen driver is defective, then the phone will have failed in its core required functionality. This is a very late stage in the life cycle to uncover such a fault.

In the waterfall model, the testing at the end serves as a quality check. The product can be accepted or rejected at this point. In the smartphone manufacturing example, this model could be adopted to check that the phone casings after manufacture are crack-free, rejecting those that have failed.

In software development, however, it is unlikely that we can simply reject the parts of the system found to be defective and release the rest. The nature of software functionality is such that removal of software is often not a clear-cut activity – this action could cause other areas to function incorrectly. It might even cause the system to become unusable. If the touchscreen driver is not functioning correctly, then some of the apps might not be accessible via a tap on the icon. On a touchscreen phone, this would be an intolerable fault in the live environment.

What is needed is a process that assures quality throughout the development life cycle. At every stage, a check should be made that the work product for that stage meets its objectives. This is a key point: work product evaluation taking place at the point where the product has been declared complete by its creator. If the work product passes its evaluation (test), we can progress to the next stage in confidence. In addition, finding problems at the point of creation should make fixing any problems cheaper than fixing them at a later stage. This is the cost escalation model, described in **Chapter 1**.

The checks throughout the life cycle include verification and validation.

Verification – checks that the work product meets the requirements set out for it. An example of this is to ensure that a website being built follows the guidelines for making websites usable by as many people as possible. Verification helps to ensure that we are building the product in the right way.

Validation – changes the focus of work product evaluation to evaluation against user needs. This means ensuring that the behaviour of the work product matches the customer needs as defined for the project. For example, for the same website above, the guidelines may have been written with people familiar with websites in mind. It may be that this website is also intended for novice users. Validation would include these users checking that they too can use the website easily. Validation helps to ensure that we are building the right product as far as the users are concerned.

There are two types of development model that facilitate early work product evaluation.

The first is an extension to the waterfall model, known as the V model. The second is a cyclical model, where the coding stage often begins once the initial user needs have been captured. Cyclical models are often referred to as iterative models.

We will consider first the V model.

V model (sequential development model)

There are many variants of the V model. One of these is shown in **Figure 2.2**.

___

Figure 2.2 The V model for software development

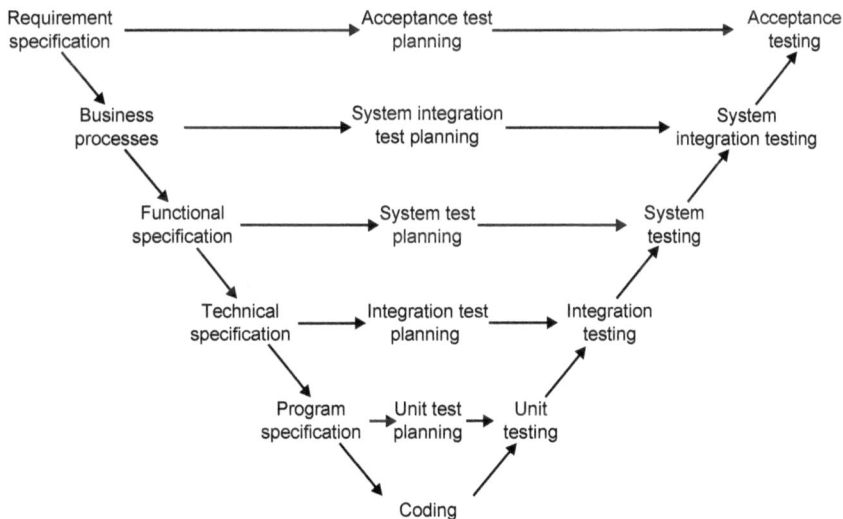

As for the waterfall model, the left-hand side of the model focuses on elaborating the initial requirements, providing successively more technical detail as the development progresses. In the model shown, these are:

- requirement specification – capturing of user needs;
- functional specification – definition of functions required to meet user needs;

- technical specification – technical design of functions identified in the functional specification;
- program specification – detailed design of each module or unit to be built to meet required functionality.

These specifications could be reviewed to check for the following:

- Conformance to the previous work product (so in the case of the functional specification, verification would include a check against the requirement specification).
- That there is sufficient detail for the subsequent work product to be built correctly (again, for the functional specification this would include a check that there is sufficient information to create the technical specification).
- That it is testable. Is the detail provided sufficient for testing the work product? Formal methods for reviewing documents are discussed in **Chapter 3**.

The middle of the V model shows that planning for testing should start with each work product. Thus, using the requirement specification as an example, acceptance testing is planned for, right at the start of development. Test planning is discussed in more detail in **Chapter 5**.

The right-hand side focuses on the testing activities. For each work product, a testing activity is identified. These are shown in **Figure 2.2**:

- Testing against the requirement specification takes place at the acceptance testing stage.
- Testing against the business processes takes place at the system integration testing stage.
- Testing against the functional specification takes place at the system testing stage.
- Testing against the technical specification takes place at the integration testing stage.
- Testing against the program specification takes place at the unit testing stage.

This allows testing to be concentrated on the detail provided in each work product, so that defects can be identified as early as possible in the life cycle, when the work product has been created. The different stages of testing are discussed later.

Each stage must be completed before the next one can be started; this approach to software development pushes validation of the system by the user representatives right to the end of the life cycle. If the customer needs were not captured accurately in the requirement specification, or if they change, then these issues may not be uncovered until the user testing is carried out. As we saw in **Chapter 1**, fixing problems at this stage could be very costly; in addition, it is possible that the project could be cancelled altogether.

The impact of this life cycle on testing includes static testing early and often, thereby reducing potential defects downstream. However, dynamic testing is then deferred to later in the development life cycle. We will explore dynamic testing techniques further in **Chapter 4**.

Iterative and incremental development models

Let us now look at a different model for software development – iterative (and typically incremental) development. This is one in which the requirements do not need to be fully defined before coding can start. Instead, a working version of the product is built, in a series of stages or iterations – hence the name iterative/incremental development. Each stage encompasses requirements definition, design, code and test. This is shown diagrammatically in **Figure 2.3**.

Figure 2.3 Iterative development

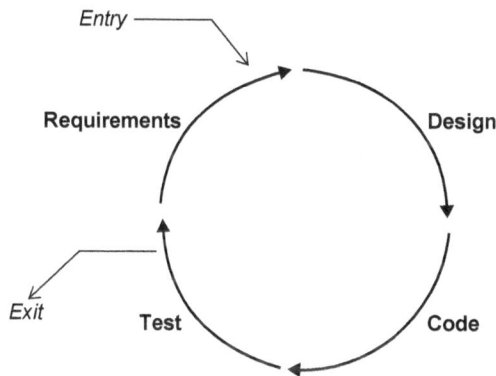

This type of development is often referred to as cyclical as we go round the development cycle a number of times within the project. The project will have a defined timescale and cost. Within this, the cycles will be defined. The cycles are commonly referred to as time-boxes. For each time-box, a requirement is defined and a version of the code is produced, which will allow testing by the user representatives. At the end of each time-box, a decision is made on what extra functionality needs to be created for the next iteration. This process is repeated until a fully working system has been produced.

Some models incorporate the idea of 'self-organising' teams. This does not mean that the team is leaderless, but rather that the team decides how to best manage and execute the tasks among themselves. This will, of course, include the relationship between testers and developers, and how defects are reported.

A key feature of this type of development is the involvement of user representatives in the testing. Having the users represented throughout minimises the risk of developing an unsatisfactory product. The user representatives are empowered to request changes to the software to meet their needs.

Components or systems developed using these methods often involve overlapping and iterating test levels throughout development. Ideally, each feature is tested at several test levels before delivery. This is often facilitated by continuous delivery or deployment, enabled by making use of significant automation.

This approach to software development can pose problems, however.

The lack of formal documentation can make it difficult to test. To counter this, developers may use test-driven development (TDD). This is where functional tests are written first, and code is then created and tested. It is reworked until it passes the tests. Both tests and code are then refactored. This is an example of using testing as a driver for software development. As we saw in **Chapter 1**, early testing saves time and money. Since tests are written before the code, this approach is often also referred to as 'shift-left'.

In addition, the working environment may be such that developers make any changes required without formally recording them. This approach could mean that changes cannot be traced back to the requirements, nor to the parts of the software that have changed. Thus, traceability as the project progresses is reduced. To mitigate this, a robust process must be put in place at the start of the project to manage these changes (often part of a configuration management process – this is discussed further in **Chapter 5**).

Another issue associated with changes is the amount of testing required to ensure that implementation of the changes does not cause unintended changes to other parts of the software (this is called regression testing, discussed later in this chapter).

Forms of iterative development include Scrum, Kanban, spiral, acceptance test-driven development (ATDD), behaviour-driven development (BDD), domain-driven development (DDD); extreme programming (XP), feature-driven development (FDD), lean IT and TDD. Agile is an umbrella term incorporating these and other methods.

- **Scrum** – the focus is on short iterations spanning just hours, days or a few weeks. The increments developed are thus correspondingly small. This term may be familiar to those of you already working in an Agile environment.

- **Kanban** – as for Scrum, you may already be familiar with this term. It allows for easy visualisation of a workflow via the usual task board used in Agile development projects. It is not a time-boxing tool; it can be used to show progress of a single enhancement or group of features, from a 'to-do' state to a 'done' state.

- **Unified Process** – increments and thus iterations here tend to be longer than in Scrum, with correspondingly larger feature sets. Those of you working in this environment may recall the Rational Unified Process: Inception–Elaboration–Construction–Transition phases.

- **Spiral** – Dr Barry Boehm is credited with this model. Here, risk is used as the driver for determining the levels of documentation and effort required for a given project. This can include a prototyping model, where increments created may be reworked significantly or even abandoned if the risks are too high.

- **ATDD** – this is another example of the test-first approach described above in TDD. Here the stakeholders (including customer representatives, developers and testers) are heavily involved in defining the criteria for accepting a user story. The creation of the code is then driven from these criteria.

- **BDD** – here the developer writes tests in a natural language based on the expected behaviour of the system (typically using the given–when–then construct). These tests are compiled and translated into automatically executable tests.

- **DDD** – this uses a Ubiquitous Language, where, as the name suggests, the same terminology is used by all stakeholders across a team. This language is also reflected at code level to further reduce possible ambiguities and increase quality across the entire life cycle. The phrase domain-driven development was coined by Eric Evans.

- **FDD** – this follows a five-step process focused on features. Create a model, create a list of features, then plan, design and build to those features.

- **XP** – introduced by Kent Beck as an Agile process incorporating values such as communication and feedback. This includes principles such as improvement and accepted responsibility, and primary practices such as whole team and continuous integration. We see these in many Agile projects today.

- **Lean IT** – as the name suggests, the process is kept lean. This means improving the value flow and eliminating wasteful practices. Lean methodologies have their roots in the manufacturing industry, particularly car manufacturing, where costs quickly increase if processes are slow and inefficient.

The impact of iterative life cycles on testing includes:

- static and dynamic testing taking place at all levels for each iteration;
- frequent delivery of increments, requiring early/frequent feedback and extensive regression testing (usually automated);
- significant test automation to keep up with changes;
- use of experience-based techniques for manual testing, requiring less planning prior to running tests than those using formal design techniques.

Agile methods of developing software have gained significant ground in recent years. Organisations across business sectors have embraced this collaborative way of working and many qualifications focusing on Agile methodologies now exist. The syllabus for this qualification does not dwell on Agile; however, for completeness of learning, a summary will now be provided.

The Agile development methodology is supported through the Agile Alliance (www.agilealliance.org). The Alliance has created an Agile manifesto with four points, supported by 12 principles. The essence of these is to espouse the value of adopting a can-do and collaborative approach to creating a product. The idea is that the development teams work closely with the business, responding to their needs at the time rather than attempting to adhere to a contract for requirements that might well need to be changed prior to the launch date. Many examples can be provided to suggest that this is a suitable way of working. Going back to our smartphone example, there are many well-known phone manufacturers who failed to move with consumer demands, costing them significant market share.

A popular framework for Agile is Scrum. Scrum is not an acronym; it was taken from the game of rugby. In rugby the team huddles to agree tactics; the ball is then passed back and forth until a sprint to the touchline is attempted. In Scrum, there is a daily stand-up meeting to agree tactics for the day; an agreed set of functions to be delivered at the end of a time-box (sprint); periodic reviews of functionality by the customer representatives; and a team retrospective to reflect on the previous sprint in order to improve in the next. In Agile, the term 'user story' is common when referring to requirements, as is the term 'backlog' when referring to a set of requirements or tasks for a particular sprint.

The ISTQB now offers qualifications in Agile testing. Further information can be found at www.istqb.org

DevOps and testing

DevOps seeks to harness the skills across the disciplines (operations, development and testing) through collaboration, flexibility and automation, with the common goal of optimising processes.

Testing benefits from DevOps include:

- fast and frequent feedback on code quality (including non-functional characteristics, such as performance and reliability) so that issues can be addressed early and cost-effectively;
- continuous integration (CI) of high-quality code by developers, with the associated component tests so that they too can deal with issues early, thus shifting left;
- the provision of stable test environments through established configuration management processes;
- significant automation across the delivery pipeline, thereby reducing the need for repetitive manual testing;
- reduction of the risks of regression by automating the regression tests where possible.

For these benefits to be realised, the following challenges should be addressed early:

- agreement of the delivery pipeline and sticking to it;
- maintaining tools and automation processes, in particular the CI tools;
- provision of suitably skilled people to carry out the test automation for the whole project.

Shift-left approach

If we think of the V model, shifting testing left means that we focus on testing the documentation first. We then test the code as it is built, followed by the typical end-to-end testing. Specific activities include:

- Reviewing the specification for omissions and ambiguities so that time is not wasted later when trying to write code or prepare/run tests.

- The developers writing tests first (as we saw in TDD) and running the code in a test harness to detect real-world problems early.

- The developers carrying out CI and delivery so that feedback can occur early. In addition, the automated tests are uploaded with the source code to the repository for future reference.

- The developers checking their source code for errors and unnecessary complexity through static analysis.

- Carrying out non-functional testing such as performance testing at the component level to detect issues before the complete system is tested, thereby reducing cost and effort.

It should be borne in mind that, as with all activities, costs are associated with the shift-left approach. These include training, infrastructure and time costs, which are incurred upfront.

Retrospectives and process improvement

Retrospectives are lessons learned from carrying out a piece of work. They help to improve future processes. They commonly involve the project stakeholders, which includes testers, having a meeting to share their views on:

- what went well, so that good practices are retained;

- what did not go so well and what needs to be improved;

- how to get the improvements into the process in a sustainable way so that the same mistakes are not repeated in the future.

These agreed improvements should be recorded, perhaps as part of the test completion report.

For testing, benefits include:

- faster creation of tests to highlight more potential defects;

- higher-quality test cases that can be related back to the user stories/requirements and find more bugs;

- higher-quality requirements and user stories to reduce time and effort in backlog refinement by the tester;

- improved communications and cooperation across the teams, especially between development and testing;

- increased overall team satisfaction.

These meetings can be held at any time, depending on the life cycle model and project constraints.

For both types of software development life cycles, testing plays a significant role. Testing helps to ensure that the work products are being developed in the right way (verification) and that the product will meet the user needs (validation).

Characteristics of good testing across the development life cycle include:

- Early test design. In the V model we saw that test planning begins with the specification documents. This activity is part of the test process, discussed in **Chapter 1**. After test planning, the documents are analysed and test cases designed. This approach ensures that testing starts with the development of the requirements; that is, a proactive approach to testing is undertaken. Proactive approaches to test design are discussed further. As we saw in iterative development, TDD may be adopted, pushing testing to the front of the development activity.

- Each work product is tested. In the V model, each document on the left is tested by an activity on the right. Each specification document is called the test basis – that is, it is the basis on which tests are created. In iterative development, the functionality for each iteration is tested before moving on to the next.

- Each test level has objectives specific to that level. Thus, at unit level the focus is on individual pieces of code; at integration level the focus is on the interfaces; and so on.

- Testers are involved in reviewing requirements before they are released. In the V model, testers are invited to review associated documents from a testing perspective.

TEST LEVELS

The test stages of the V model are shown in **Figure 2.2**. They are often called test levels. The term test level provides an indication of the focus of the testing, and the types of problems it is likely to uncover. The typical levels of testing are:

- component (unit) testing;
- integration testing;
- system testing;
- system integration testing;
- acceptance testing.

Each of these test levels will include tests designed to uncover problems specific to that stage of development. These levels of testing can also be applied to iterative development. The levels may change depending on the system. For instance, if the system includes some software developed by external parties, or bought off the shelf (commercial off-the-shelf (COTS)), acceptance testing on these may be conducted before testing the system as a whole.

Each test level will have a test basis (a description of the item) and a test object (the item under test). A test basis is some form of definition of what the code is intended to do and is used as a reference for deriving the tests. It can include the requirements, user stories, the source code or the knowledge of the tester, based on experience. The higher the level of detail provided in the documentation, the more precise the test design can be. Typically there is more documentation in V model development than in iterative development. Techniques for test design will be covered in **Chapter 4**.

Test levels are characterised by the following attributes:

- specific objectives;
- test basis, referenced to derive test cases;
- test object (i.e. what is being tested);
- typical defects and failures;
- specific approaches and responsibilities.

Here we look at these levels of testing in more detail.

Component (unit) testing

Before testing of the code can start, clearly the code has to be written. This is shown at the bottom of the V model and as part of the cycle in iterative development. Generally, the code is written in component parts, or units. The components are usually constructed in isolation, for integration at a later stage. Components are also units.

Component (unit) testing, normally performed by developers, is often done in isolation from the rest of the system, depending on the SDLC model and the system. This typically requires mock objects, service virtualisation, harnesses, stubs and drivers. These become necessary when actual code for components is not available, allowing the developer to check the functionality of their own piece of code fully.

Component (unit) testing may cover:

- Functional requirements, such as the ability to remove items from a shopping cart.
- Non-functional characteristics, such as checking for memory leaks (this is where the program holds on to memory it is no longer using, which may cause the system to slow down when in use).
- Structural testing – this is checking the percentage of code exercised through testing. Testing based on code (white box testing) is discussed in **Chapter 4**.

Integration testing

Once the units have been written, the next stage is to put them together to create the system. This is called integration. It involves building something larger from a number of smaller pieces.

The purpose of integration testing is to expose defects in the interfaces and in the interactions between integrated components or systems.

There are two different levels of integration testing described in the ISTQB syllabus, which may be carried out on test objects of varying size as follows:

- Component (unit) integration testing, which focuses on the interactions and interfaces between integrated components. It is performed after component testing and is generally automated. In iterative and incremental development, component integration tests are usually part of the CI process. As might be expected, the integration tests need to be planned to match the order in which the components will be built so that the units of code are available to be integrated. This is often referred to as the integration strategy. This can include a bottom-up strategy, where the approach to integrating and testing the units of code is to integrate the code one unit at a time and then testing before adding the next unit of code.

- System integration testing, which focuses on the interactions and interfaces between systems. It can also cover interactions and interfaces with external organisations. For example, a trading system in an investment bank may interact with the stock exchange to get the latest prices for its stocks and shares on the international market. Where external organisations are involved, extra challenges for testing present themselves, since the developing organisation will not have control over the interfaces. This can include creating the test environment, defect resolution and so on.

Systematic integration strategies may be based on the system architecture (e.g. top-down and bottom-up), functional tasks, transaction processing sequences or some other aspect of the system or components.

There are three commonly quoted integration strategies, as follows.

Big-bang integration
This is where all units are linked at once, resulting in a complete system. When the testing of this system is conducted, it is difficult to isolate any errors found because attention is not paid to verifying the interfaces across individual units.

This type of integration is generally regarded as a poor choice of integration strategy. It introduces the risk that problems may be discovered late in the project, when they are more expensive to fix.

Top-down integration
This is where the system is built in stages, starting with components that, when activated, cause other components to become active. These are called 'calling' components. Components that call others are usually placed above those that are called. Top-down integration testing permits the tester to evaluate component interfaces, starting with those at the 'top'.

Let us look at the diagram in **Figure 2.4** to explain this further.

The control structure of a program can be represented in a chart. In **Figure 2.4**, component 1 can call components 2 and 3. Thus, component 1 is placed above components 2 and 3. Component 2 can call components 4 and 5. Component 3 can call components 6 and 7. Thus, components 2 and 3 are placed above components 4 and 5 and components 6 and 7, respectively.

Figure 2.4 Top-down control structure

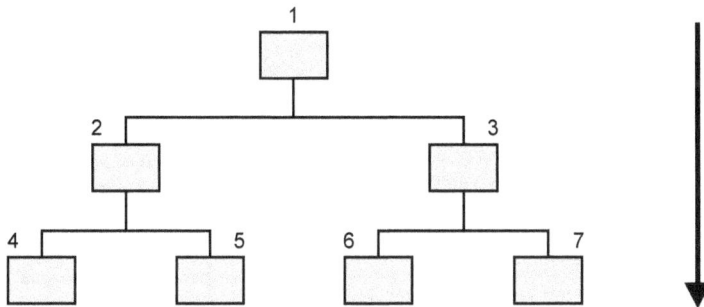

In this chart, the order of integration might be:

- 1,2
- 1,3
- 2,4
- 2,5
- 3,6
- 3,7.

Top-down integration testing requires that the interactions of each component must be tested when they are built. Those lower down in the hierarchy may not have been built or integrated yet. In **Figure 2.4**, in order to test component 1's interaction with component 2, it may be necessary to replace component 2 with a substitute since component 2 may not have been integrated yet. This is done by creating a skeletal implementation of the component, called a stub. A stub is a passive component that is called by other components. In this example, stubs may be used to replace components 4 and 5 when testing component 2.

The use of stubs is commonplace in top-down integration, replacing components not yet integrated.

Bottom-up integration
This is the opposite of top-down integration. The components are integrated in a bottom-up order, as shown in **Figure 2.5**.

The integration order might be:

- 4,2
- 5,2
- 6,3
- 7,3
- 2,1
- 3,1.

Figure 2.5 Bottom-up integration

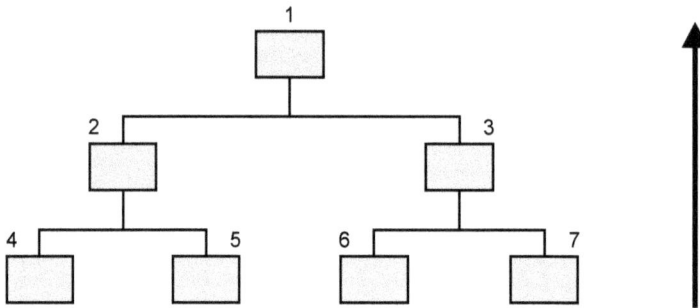

So, in bottom-up integration components 4–7 are integrated before components 2 and 3. In this case, the components that may not be in place are those that actively call other components. As in top-down integration testing, they must be replaced by specially written components. When these special components call other components, they are called drivers. They are so called because, in the functioning program, they are active, controlling other components. Components 2 and 3 could be replaced by drivers when testing components 4–7. They are generally more complex than stubs.

System testing

Having checked that the components all work together at unit integration level, the next step is to consider the functionality from an end-to-end perspective. This activity is called system testing.

System testing is necessary because many of the criteria for test selection at unit and integration testing result in the production of a set of test cases that are unrepresentative of the operating conditions in the live environment. Thus, testing at these levels is unlikely to reveal failures due to interactions across the whole system, or those due to environmental issues.

System testing serves to correct this imbalance by focusing on the behaviour of the whole system/product as defined by the scope of a development project or programme, in a representative live environment. It is usually carried out by a team that is independent of the development process. The benefit of this independence is that an objective assessment of the system can be made, based on the specifications as written and not the code.

System testing often produces information that is used by stakeholders to make release decisions, which may include checking that legal or regulatory requirements and standards have been met.

The behaviour required of the system may be documented in functional specifications, use cases or user stories. These should include the functional and non-functional requirements of the system or feature.

A functional requirement is a requirement that specifies a function that a system or system component must perform. Functional requirements can be specific to a system. For instance, you expect to be able to search for flights on a travel agent's website, whereas you visit your online bank to check that you have sufficient funds to pay for the flight. Thus, functional requirements provide detail on what the application being developed will do.

Non-functional system testing looks at those aspects that are important but not always directly related to what functions the system performs. These tend to be generic requirements that can be applied to many different systems. In the example above, you can expect that both systems will respond to your inputs in a reasonable time frame, for instance. Typically, these requirements will consider both normal operations and behaviour under exceptional circumstances. Thus, non-functional requirements detail how the application will perform in use.

Examples of non-functional requirements include:

- installability – installation procedures;
- maintainability – ability to introduce changes to the system;
- performance efficiency – expected normal behaviour;
- load handling – behaviour of the system under increasing load;
- stress handling – behaviour at the upper limits of system capability;
- portability – use on different operating platforms;
- recovery – recovery procedures on failure;
- reliability – ability of the software to perform its required functions over time;
- usability – ease with which users can engage with the system;
- security – the level to which security systems must be in place.

The amount of testing required at system testing, however, can be influenced by the amount of testing carried out (if any) at the previous stage. In addition, the amount of testing advisable may also depend on the amount of verification carried out on the requirements (this is discussed further in **Chapter 3**).

Acceptance testing

The purpose of acceptance testing is to provide the end users with confidence that the system will function according to their expectations.

Unlike system testing, however, the testing conducted here should be independent of any other testing carried out. Its key purpose is to demonstrate system conformance to, for example, the customer requirements and operational and maintenance processes. For instance, acceptance testing may assess the system's readiness for deployment and use.

Typical forms of acceptance testing include the following:

- **User acceptance testing** – testing by user representatives to check that the system meets their business needs. This can include factory acceptance testing, where the system is tested by the users before moving it to their own site. Site acceptance testing could then be performed by the users at their own site.

- **Operational acceptance testing** – often called operational readiness testing. This involves checking that the processes and procedures are in place to allow the system to be used and maintained. This can include checking:
  - back-up facilities;
  - installing, uninstalling and upgrading;
  - performance issues;
  - procedures for disaster recovery;
  - user management;
  - maintenance procedures;
  - data load and migration tasks;
  - security vulnerabilities.

- **Contract and regulatory acceptance testing:**
  - Contractual acceptance testing – sometimes the criteria for accepting a system are documented in a contract. Testing is then conducted to check that these criteria have been met, before the system is accepted. This is typical of custom-developed software.
  - Regulatory acceptance testing – in some industries, systems must meet governmental, legal or safety standards. Examples of these are the defence, banking and pharmaceutical industries. The results of tests here may be witnessed or audited by regulatory bodies.

- **Alpha and beta testing:**
  - Alpha testing takes place at the developer's site – the operational system is tested while still at the developer's site by people outside the organisation before being released to external customers. Note that testing here is still independent of the development team.
  - Beta testing takes place at the customer's site – the operational system is tested by a group of customers who use the product at their own locations and provide feedback before the system is released. This is often called 'field testing'.

Both alpha and beta testing are typically used by developers of COTS software in order to get feedback before final go-live.

In iterative development, project teams can employ various forms of acceptance testing during and at the end of each iteration, such as those focused on verifying a new feature against its acceptance criteria and those focused on validating that a new feature satisfies the users' needs. In addition, alpha tests and beta tests may occur, either at the end of each iteration, after the completion of each iteration or after a series of iterations. User acceptance tests, operational acceptance tests, regulatory acceptance tests and contractual acceptance tests also may occur, either at the close of each iteration, after the completion of each iteration or after a series of iterations.

TEST TYPES

In the previous section we saw that each test level has specific testing objectives. In this section we will look at the types of testing required to meet these objectives.

Test types fall into the following categories:

- functional testing;
- non-functional testing;
- white box testing.

All of these can be used at any level of testing and indeed can be used when reviewing documents as part of static testing.

Functional testing

As you saw in the section on system testing, functional testing looks at the specific functionality of a system, such as searching for flights on a website, or perhaps calculating employee pay correctly using a payroll system. This can include checks for completeness, correctness and appropriateness.

Functional testing is carried out at all levels of testing, from unit through to acceptance testing. In the example above, the testing of the correct calculation of employee pay may be done during unit testing, whereas searching for a flight is often done during system testing.

Functional testing is also called specification-based testing or black box testing (covered in **Chapter 4**). It can be measured in terms of the percentage of requirements covered by the tests.

Designing tests at this level often requires specific domain skills. In today's world, testing the blockchain used in cryptocurrencies requires different knowledge and skills to those used in designing tests for normal banking operations, for instance.

Non-functional testing

This is where the behavioural aspects of the system are tested. As you saw in the section on system testing, examples include usability, performance, efficiency and security testing, among others.

As for functional testing, non-functional testing:

- should be performed at all levels so that potential defects are detected as early as possible, since late detection can be very serious and expensive to fix;
- can make use of black box testing techniques, such as checking that a flight can be booked within a specific time frame;
- often requires specialist knowledge (such as knowing the inherent weaknesses of specific technologies), skills (such as having an understanding of how to carry out performance testing) and test environments (such as usability labs);
- can be measured – for instance, checking the percentage of mobile devices tested for compatibility with an application.

These tests can be referenced against a quality model, such as the one defined in ISO/IEC 25010 *Systems and Software: Quality Requirements and Evaluation (SQuaRE)*. Although a detailed understanding of this standard is not required for the exam, we provide a brief summary here. As a part of non-functional testing, we can check the following:

- Performance efficiency – including response times, resource utilisation and maximum limits of performance. You are probably familiar with websites with lots of graphics that take longer to load than other sites that you might visit.
- Compatibility – information-sharing between products and systems within a common environment, such as buying a product online. This requires a front end to allow a user to make a purchase and a back end to process payment and arrange delivery of the items.
- Usability – this we are familiar with. Does the system provide what we need efficiently and effectively? In today's world of artificial intelligence, this aspect of testing opens up a potential divide between efficiency (machine learning algorithms run fast) and effectiveness (will the desired human outcomes match the determination of the machine?).
- Reliability – does the system perform its functions under normal operating conditions and can it cope with hardware or software faults? For instance, if a credit card payment has not been successful, is the user reliably informed of this and how to put it right?
- Security – in recent years we have been made aware of the many security risks associated with online transactions. For example, banks have responded in an attempt to stop phishing by taking the user through a series of questions before a payment can be made.
- Maintainability – for many of us, once a system goes live we move on to the next project. Maintainability looks at the ease of making changes to the system to respond to possible live defects, changes to the environment or shifting user demands, for example.

- Portability – this looks at the ability to adapt, replace and install systems in different hardware and software environments. For example, iPhone™ users would be familiar with iOS updates, which should be seamless to them.

White box testing

In white box testing our focus is on the internal structure of the system. This could be the code itself, an architectural definition or data flows through the system.

White box testing is commonly carried out at unit and component integration test levels. Here, common measures include code and interface coverage (percentage of code and interfaces exercised by tests). Further detail on code coverage measures is provided in **Chapter 4**.

It can also be carried out at the higher levels of testing where a structural definition of the system exists. An example is a business flow (represented as a flow chart), which could be used to design tests at system or higher levels.

As before, this type of testing also requires specialised knowledge and skills, such as code creation, data storage on databases and use of the associated tools.

CONFIRMATION AND REGRESSION TESTING

The previous sections detail the testing to be carried out at the different stages in the development life cycle. At any level of testing, it can be expected that defects will be discovered. When these are found and fixed, the quality of the system being delivered is improved.

After a defect is detected and fixed, the changed software should be retested to confirm that the problem has been successfully removed. This can be checked by rerunning the test cases that highlighted the problem, as well as adding new tests for any changes made to fix the defect. This is called retesting or confirmation testing.

Note that when the developer removes the defect, this activity is called debugging, which is not a testing activity. Testing finds a failure, debugging fixes it.

The unchanged software should also be retested to ensure that no additional defects have been introduced as a result of changes to the software. This is called regression testing. Regression testing should also be carried out if the environment has changed.

Regression testing involves the creation of a set of tests which serve to demonstrate that the system works as expected. These are run many times over a testing project, when changes are made, as discussed above. This repetition of tests makes regression testing suitable for automation in many cases. This is particularly true when automated builds and CI, such as in DevOps, are used. Test automation is covered in detail in **Chapter 6**.

In iterative development projects such as Agile development, the requirements churn introduces a great need for both confirmation and regression testing. There is also a concept of code refactoring (where a developer seeks to increase the quality of the code written), which also necessitates change-related testing.

CHECK OF UNDERSTANDING

Which of the following is correct?

A. Regression testing checks that a problem has been successfully addressed, while confirmation testing is done at the end of each release.

B. Regression testing checks that all problems have been successfully addressed, while confirmation testing refers to testing individual fixes.

C. Regression testing checks that fixes to errors do not introduce unexpected functionality into the system, while confirmation testing checks that fixes have been successful.

D. Regression testing checks that all required testing has been carried out, while confirmation testing checks that each test is complete.

MAINTENANCE TESTING

For many projects (though not all) the system is eventually released into the live environment. Hopefully, once deployed, it will be in service as long as intended, perhaps for years or decades.

During this deployment it may become necessary to change the system.

Triggers for maintenance fall into three broad categories: modifications, upgrades or migrations and retirement. This can include:

* additional features being required;
* the system being migrated to a new operating platform;
* the system being retired – data may need to be migrated or archived;
* planned upgrade to COTS-based systems;
* new faults being found requiring fixing (these can be 'hot fixes').

Once changes have been made to the system, they will need to be tested (retesting), and it also will be necessary to conduct regression testing to ensure that the rest of the system has not been adversely affected by the changes. Testing that takes place on a system that is in operation in the live environment is called maintenance testing.

When changes are made to migrate from one platform to another, the system should also be tested in its new environment. When migration includes data being transferred in from another application, then conversion testing also becomes necessary.

As we have suggested, all changes must be tested, and, ideally, all of the system should be subject to regression testing. In practice this may not be feasible or cost-effective. An understanding of the parts of the system that could be affected by the changes could reduce the amount of regression testing required. Working this out is termed impact analysis; that is, analysing the impact of the changes on the system.

Impact analysis for maintenance

The purpose of impact analysis is to determine the likely impact of a change to a system. We need to understand the intentions of the change, any potential side effects of the change and how existing tests may need to be changed. This may result in the change not being made at all if the costs outweigh the benefits.

This can be difficult for a system that has already been released and is in maintenance. This is because the specifications may be out of date (or non-existent); test cases may not have been documented; there is a lack of traceability of tests back to requirements; there is weak or non-existent tool support; or the original development team may have moved on to other projects or left the organisation altogether.

Thus, it is important that live systems maintain the test basis (e.g. requirement specifications), the testware (e.g. test cases) and importantly the traceability between the two, otherwise impact analysis may become very difficult to achieve.

CHECK OF UNDERSTANDING

1. How do functional requirements differ from non-functional requirements?
2. For which type of testing is code coverage measured?
3. What is the purpose of maintenance testing?
4. Give examples of when maintenance testing is necessary.
5. What is meant by the term impact analysis?

SUMMARY

In this chapter we have explored the role of testing within the SDLC. We have looked at the basic steps in any development model, from understanding customer needs to delivery of the final product. These were built up into formally recognisable models, using distinct approaches to software development.

The V model, as we have seen, is a stepwise approach to software development, meaning that each stage in the model must be completed before the next stage can be started if a strict implementation of the model is required. This is often the case in safety-critical developments. The V model typically has the following work products and activities:

1. requirement specification;
2. functional specification;
3. technical specification;

4. program specification;

5. code;

6. unit testing;

7. integration testing;

8. system testing;

9. system integration testing

10. acceptance testing.

Work products 1–5 are subject to verification, to ensure that they have been created following the rules set out. For example, the program specification is assessed to ensure that it meets the requirements set out in the technical specification, and that it contains sufficient detail for the code to be produced.

In activities 6–10 the code is assessed progressively for compliance to user needs, as captured in the specifications for each level.

An iterative model for development has fewer steps but involves the user from the start. These steps are typically:

1. define iteration requirement;

2. build iteration;

3. test iteration.

This sequence is repeated for each iteration until an acceptable product has been developed.

An explanation of each of the test levels in the V model was given. For unit testing the focus is the code within the unit itself; for integration testing it is the interfacing between units and systems; for system testing it is the end-to-end functionality; and for acceptance testing it is the user perspective.

An explanation of test types was then given, and by combining test types with test levels we can construct a test approach that matches a given system and a given set of test objectives very closely. The techniques associated with test types are covered in detail in **Chapter 4**, and the creation of a test approach is covered in **Chapter 5**.

Finally, we looked at the testing required when a system has been released but a change has become necessary – maintenance testing. We discussed the need for impact analysis in deciding how much testing to do after the changes have been implemented. This can pose an added challenge if the requirements associated with the system are missing or have been poorly defined.

In the next chapter, techniques for improving requirements will be discussed.

Example examination questions with answers

E1. K1 question
Which of the following are test levels where white box testing is applicable?

    i.   Unit testing.
    ii. Acceptance testing.
    iii. Regression testing.
    iv. Performance testing.

    A. i and ii.
    B. i only.
    C. ii and iii.
    D. ii and iv.

E2. K2 question
Which of the following is true of non-functional testing?

    A. Examples of non-functional testing are provided in ISO standard 20246.
    B. It is best carried out at system and acceptance test levels.
    C. It cannot usually be measured.
    D. It can make use of black box test techniques.

E3. K2 question
Which of the following approaches to iterative development models is considered test-first?

    A. Kanban.
    B. Domain-driven development.
    C. Behaviour-driven development.
    D. Scrum.

E4. K2 question
Which of the following statements are examples of good testing practices?

    i.   For every development activity there is a corresponding testing activity.
    ii. Component testing is reduced in iterative development since the users will do most of the testing.
    iii. Confirmation and regression testing are left to the system and acceptance testers so that the developers can focus on coding.
    iv. Testers should be involved in reviewing documents as soon as drafts are available in the development life cycle.

    A. i and ii.
    B. iii and iv.
    C. ii and iii.
    D. i and iv.

E5. K2 question
Which of the following is not true of regression testing?

A. It can be carried out at each stage of the life cycle.

B. It serves to demonstrate that the changed software works as intended.

C. It serves to demonstrate that software has not been unintentionally changed.

D. It is often automated.

Answers to the self-assessment questions in the chapter

SA1. The correct answer is D.

SA2. The correct answer is B.

SA3. The correct answer is A.

Answers to example examination questions

E1. The correct answer is A.

White box testing is applicable at all test levels. Regression and performance testing are not test levels; they are test types.

E2. The correct answer is D.

- A. provides a standard for use in reviews. The standard used for non-functional testing is ISO 25010.
- B. is incorrect – non-functional testing should be carried out all levels.
- C. is incorrect, it can be measured in terms of percentage of non- functional requirements covered.

E3. The correct answer is C.

- A. (Kanban) is about visualisation of the workflow across the team.
- B. (DDD) is about use of a common language across a team.
- D. (Scrum) is a framework for the whole team.

E4. The correct answer is D.

Option ii is incorrect – each test level has a different objective. Option iii is also incorrect – test analysis and design should start once the documentation has been completed.

E5. The correct answer is B.

This is a definition of confirmation testing. The other three options are true of regression testing.

3 STATIC TESTING

Geoff Thompson

INTRODUCTION

This chapter introduces an important area of software testing – static testing. Static testing techniques test software work products and software without executing it. They are important because they can find errors and defects before code is built/executed and therefore earlier in the life cycle of a project, making corrections easier and cheaper to achieve than for the same defects found during test execution. Review types and techniques are central to the static testing approach, and in this chapter we will look at the alternative types of review and how they fit with the test process that was defined in **Chapter 1**.

Learning objectives

The learning objectives for this chapter are listed below. You can confirm that you have achieved these by using the self-assessment questions immediately following the learning objectives, the 'Check of understanding' boxes distributed throughout the text and the example examination questions provided at the end of the chapter. The chapter summary will remind you of the key ideas.

Each learning objective is allocated a K number to represent the level of understanding required; see the **Introduction** (pp. 2–3) for an explanation of K numbers.

Static testing basics

- FL-3.1.1 (K1) Recognise types of products that can be examined by the different static test techniques.
- FL-3.1.2 (K2) Explain the value of static testing.
- FL-3.1.3 (K2) Compare and contrast static and dynamic testing.

Feedback and review process

- FL-3.2.1 (K1) Identify the benefits of early and frequent stakeholder feedback.
- FL-3.2.2 (K2) Summarise the activities of the review process.

- FL-3.2.3 (K1) Recall which responsibilities are assigned to the principal roles when performing reviews.

- FL-3.2.4 (K2) Compare and contrast the different review types.

- FL-3.2.5 (K1) Recall the factors that contribute to a successful review.

Self-assessment questions

The following questions have been designed to enable you to check your current level of understanding of the topics in this chapter. The answers are at the end of the chapter.

Question SA1 (K1)
One of the roles in a review is that of facilitator. Which of the following best describes this role?

    A.   Ensures the effective running of the review meetings, where it is decided that a meeting is required.

    B.   Allocates time in the plan, decides which reviews will take place and ensures that the benefits are delivered.

    C.   Writes the document to be reviewed, agrees that the document can be reviewed and updates the document with any changes.

    D.   Documents all issues raised in the review meeting, records problems and open points.

Question SA2 (K2)
Which of the following statements are correct for walkthroughs?

    i.   Often led by the author.

    ii.   Documented and defined results.

    iii.   All participants have defined roles.

    iv.   Used to aid learning.

    v.   Main purpose is to find defects.

    A.   i, ii and v are correct.

    B.   ii, iii and iv are correct.

    C.   i, iv and v are correct.

    D.   iii, iv and v are correct.

Question SA3 (K1)
Which of the following is an activity in the review process?

    A.   Design of the document.

    B.   Booking meeting rooms.

    C.   Writing program code.

    D.   Fixing and reporting.

BACKGROUND TO STATIC TESTING

Static testing tests software and work products without executing any code. Typically, this includes requirement specification documentation, code, process specification, system architecture specification and other work products. The first of these static test techniques is known as a review and is typically used to find and remove errors and ambiguities in documents before they are used in the development process, thus reducing one source of defects in the code. The second is known as static analysis, and it enables code to be analysed for structural defects or systematic programming weaknesses that may lead to defects without executing the code.

Static techniques find the causes of failures rather than the failure itself, which is found during test execution.

Reviews are normally completed manually; static analysis is normally completed automatically using tools.

Giving a draft document to a colleague to read is the simplest example of a review, and one that can yield a larger crop of errors than often anticipated.

WORK PRODUCTS THAT CAN BE EXAMINED BY STATIC TESTING

Virtually any work product can be examined using static testing techniques. A work product is anything that is written down. It can include:

- specifications such as business requirements, functional requirements, security requirements and non-functional requirements;
- epics, user stories and acceptance criteria;
- architecture and design specifications;
- testware such as test plans and test cases;
- user guides;
- web pages;
- contracts, plans, schedules and budgets;
- design models;
- code – however, this would need a structure against which it can be checked (e.g. models, code or text with a formal syntax).

Some work products are not appropriate for static testing. This includes those that are difficult to interpret by human beings, such as large artificial intelligence databases, and work products that should not be analysed by tools such as third-party executable code, where access is locked for legal reasons.

VALUE OF STATIC TESTING

The earlier in the life cycle that static testing is applied, the larger the benefits are. If requirement specifications or design documentation is statically tested before any code is written, this will remove defects from the work product and ensure these defects are not built into the code. As was shown in **Chapter 1**, the earlier a defect is found, the cheaper it is to fix. From the example in **Chapter 1** we can see that if a defect in the design documentation is built into code and even deployed to the user base, the fix can be enormously expensive.

How much more expensive a defect found in live use could be can be seen in the following example: A bank develops a letter to inform their clients of a change in interest rates. A small defect (spelling mistake), if found in the initial letter design before the letter is coded, costs no more than a few seconds of the author's time to correct. However, if that spelling mistake moves from the design into the coded letter and goes into live use without being spotted, the cost to fix it increases considerably. The letter needs to be taken out of live use, corrected and retested, which could take up to a week to complete, rather than the few seconds it would take to fix if found earlier.

Static testing benefits may include:

- detecting and correcting defects more efficiently, and prior to test execution;
- identification of defects not easily found by dynamic testing;
- preventing defects in design or coding by uncovering inconsistencies, ambiguities, contradictions, omissions, inaccuracies and redundancies in requirements;
- increasing development productivity;
- reducing development cost and time;
- reducing dynamic testing cost and time;
- reducing total cost of quality over the software's lifetime due to fewer failures later in the life cycle or after delivery into live operation;
- improving communication between team members in the course of participating in reviews.

Benefits recognition is key to ensuring that static testing remains a focus, as when no one recognises any benefits there is a better than average chance that static testing will be removed from the project plans. It is important, therefore, that the results of static testing are made clear to all stakeholders regularly.

DIFFERENCES BETWEEN STATIC AND DYNAMIC TESTING

Static and dynamic testing have a similar objective: to find defects as soon as possible. The main difference is that static testing is carried out against work products without executing any code, whereas dynamic testing is carried out by executing actual code or the final software or hardware product. Static testing is most effective when it is used to find defects before any code is written, thereby ensuring that the code that is written is not based on wrong or faulty specifications and so on.

They may have the same objective, but there are some differences in how they those objectives are achieved:

- Static testing is about the prevention and detection of defects, whereas dynamic testing is about finding defects.

- There are some defect types that can only be found by either static testing (undeclared variables, boundary violations, syntax violations, inconsistent interface) or dynamic testing (software behaviour, system performance).

- Static testing finds defects directly in the work products, while dynamic testing causes failures from which the associated defects are determined through subsequent analysis.

- Often, code may not be exercised that often, or is deeply embedded, so building a dynamic test to find defects is sometimes too hard or impossible to do. In this situation static testing can be used to detect defects that are deeply embedded in the code.

- Static testing can be applied to non-executable work products, while dynamic testing can only be applied to executable work products.

- Static testing can be used to measure quality characteristics that are not dependent on executing code (e.g. maintainability, such as poor reuse of components, code that is difficult to analyse and modify without introducing new defects, improper modularisation or no documentation at all), while dynamic testing can be used to measure quality characteristics that are dependent on executing code (e.g. performance efficiency).

Typical defects that are easier and cheaper to find during static testing include:

- requirements defects (e.g. inconsistencies, ambiguities, contradictions, omissions, inaccuracies and redundancies);

- design defects (inefficient algorithms or database structures), high coupling (the lack of interdependence between software modules) and low cohesion (associated with undesirable traits such as being difficult to maintain, test, reuse or even understand);

- coding defects (e.g. variables with undefined values, variables that are declared but never used, unreachable code and duplicate code);

- deviations from standards (e.g. lack of adherence to coding standards);

- incorrect interface specifications (e.g. different units of measurement used by the calling system than by the called system);

- security vulnerabilities (e.g. susceptibility to buffer overflows);

- gaps or inaccuracies in test basis traceability or coverage (e.g. missing tests for an acceptance criterion).

FEEDBACK AND REVIEW PROCESS

The review process helps to achieve early and frequent feedback on the quality of the work products. Involving key stakeholders early in the process reviewing requirements, for example, will ensure defects in the requirements are identified long before they become code, but more importantly ensure the stakeholder is clear what product and design they will be receiving.

Involving the stakeholder is key to ensuring that the product being developed absolutely meets their requirements and vision. Any failure to deliver what the stakeholder wants will lead to project delays, long debates on what is right and wrong, costly rework, deadlines missed and finger pointing (the blame game). In the worst case it may lead to complete project failure.

It is important that this review feedback is sought throughout the software development life cycle (SDLC) to ensure that any misunderstandings regarding requirements are clarified, and that any changes to requirements are fully understood and implemented correctly at the earliest point. The stakeholder feedback really helps the development team to improve their understanding of what they have been instructed to build, ensuring they focus on the features that deliver the most value to the stakeholders and users.

This early feedback can also inform the risk-based test process of areas that may be more critical or complex, both cases possibly leading to the wrong solution being built. It is key that in a prioritised test approach these areas are a focus for all test activities.

ISO/IEC 20246: REVIEW PROCESS ACTIVITIES

There is an ISO/IEC standard that covers the reviews process. Its formal title is ISO/IEC 20246 *Software and Systems Engineering – Work Product Reviews*. This standard defines what it calls a generic review process for reviews such as inspections, reviews and walkthroughs that can be referenced and used by all organisations involved in the management, development, test and maintenance of systems and software. The process framework is structured but enables flexibility, allowing the user to define a specific review process that works best for their situation. The more formal a review is, the more of the described tasks will be needed.

What drives the need for a formal review?

There are two types of review, informal and formal.

An informal review is identified by the principle that it is not following a defined process and has no formal documented output.

A formal review is identified by team participation, documented results and documented procedures. Often there will also be defined roles.

The decision on the appropriate level of formality for a review is usually based on combinations of the following factors:

81

- The type of SDLC (there are different approaches to reviews in an Agile project compared to a waterfall project).

- The maturity of the development process. The more mature the process is, the more formal reviews tend to be.

- The complexity of the work product.

- Legal or regulatory requirements. These are used to govern the software development activities in certain industries, notably in safety-critical areas such as railway signalling, and determine what kinds of review should take place.

- The need for an audit trail. Formal review processes ensure that it is possible to trace backwards throughout the SDLC. The level of formality in the types of review used can help to increase how much is contained within the audit trail.

Reviews can also have a variety of objectives, where the term 'review objective' identifies the main focus for a review. Typical review objectives include:

- finding defects;
- gaining understanding;
- generating discussion;
- education;
- decision making by consensus.

The way a review is conducted will depend on what its specific objective is, so a review aimed primarily at finding defects will be quite different from one aimed at gaining understanding of a document.

All reviews, formal and informal alike, exhibit the same basic elements of process:

- The document under review is studied by the reviewers.

- Reviewers identify issues or problems and inform the author either verbally or in a documented form, which might be as formal as raising a defect report or as informal as annotating the document under review.

- The author decides on any action to take in response to the comments, and updates the document accordingly.

This basic process is always present, but in more formal reviews it is elaborated to include additional stages and more attention to documentation and measurement.

Some of the work products to be reviewed can be very large and so may be included in many review cycles and not just a single review. The review process may be invoked many times until the work product has been fully reviewed (e.g. reviewing it chapter by chapter).

Using ISO/IEC 20246 we can see that reviews all follow the same basic process, and the more formal the review, the more formal the process surrounding the review is. In the following the key stages are explained in more detail. Also see **Figure 3.1** for the stages of a review.

Figure 3.1 Stages of a review

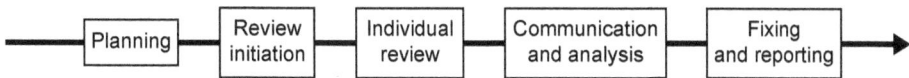

```
┌──────────┐   ┌────────────┐   ┌────────────┐   ┌───────────────┐   ┌───────────────┐
│ Planning │──▶│  Review    │──▶│ Individual │──▶│ Communication │──▶│    Fixing     │──▶
│          │   │ initiation │   │   review   │   │  and analysis │   │ and reporting │
└──────────┘   └────────────┘   └────────────┘   └───────────────┘   └───────────────┘
```

Planning

- Defining the scope – this includes the purpose of the review; for example, finding defects, what documents or parts of documents are to be reviewed and the quality characteristics to be evaluated.

- Estimating the effort required and the time frame that the review will be undertaken in – much like the estimating required to fund the test activity, reviews need to be estimated so they can be costed in a project budget.

- Identifying the review characteristics, such as review type, roles, activities and checklists – as described above, there are many factors that help the decision (e.g. what type of review characteristics will be used).

- Selecting the people to undertake the review – ensuring that those selected can and will add value to the process. There is little point in selecting a reviewer who will agree with everything written by the author without question. As a rule of thumb, it is best to include some reviewers from a different part of the organisation, who are known to be 'picky' and known to be dissenters. Reviews, like weddings, are enhanced by including 'something old, something new, something borrowed, something blue'. In this case, 'something old' would be an experienced practitioner; 'something new' would be a new or inexperienced team member; 'something borrowed' would be someone from a different team; and 'something blue' would be the dissenter who is hard to please. At the earliest stage of the process a moderator (review leader) must be identified. This is the person who will coordinate all of the review activity. This also includes allocating roles.

- Allocating roles – each reviewer is given a role to provide them with a unique focus on the document under review. Someone in a tester role might be checking for testability and clarity of definition, while someone in a user role might look for simplicity and a clear link to business values. This approach ensures that, although all reviewers are working on the same document, each individual is looking at it from a different perspective.

- Defining the entry and exit criteria, especially for the most formal review types (e.g. inspections) – before a review can start, certain criteria have to be met. These are defined by the moderator and could include: all reviewers must have received the review papers; all reviewers have a kick-off meeting booked in their diaries; any training of reviewers has been completed.

- Checking entry criteria (mainly used for more formal review types such as inspections) – this stage is where the entry criteria agreed earlier are checked to ensure they have been met so that the review can continue.

Review initiation

The focus of this stage of the review is to ensure that everyone and everything needed for a successful review are prepared ready to start the review. This will include:

- Distributing the work products to all reviewers for review.

- Making sure everyone understands their roles by explaining the scope, objectives, process, roles and responsibilities and work products to the participants – this can be run as a meeting or simply by sending out the details to the reviewers. The work products may be described by the author. The method used will depend on timescales and the volume of information to pass on. A lot of information can be disseminated better by a meeting rather than expecting reviewers to read pages of text.

- Answering any questions raised by the review team.

- Any required training being arranged.

Individual review

- The focus of this stage is ensuring that the reviewer reviews all parts of the work product, including any source documents.

- Noting anomalies/issues, questions, recommendations and comments – in this stage the potential defects, questions and comments found during individual preparation are logged.

Communication and analysis

- Communicating potential anomalies/issues found either in a review meeting or via a log.

- Analysing the anomalies to identify actual defects; if it is agreed that there is a defect, logging this correctly, assigning ownership of the repair work, actions to be taken and the status of the defect.

- The quality characteristics of the work product are reviewed and then evaluated, along with all relevant criteria and used to make the review decision.

Fixing and reporting

- Creating defect reports for those findings that require changes, which may include making recommendations regarding handling the defects, making decisions about the defects themselves and so on.

- Fixing defects found – here, typically, the author will fix defects that were found and agreed as requiring a fix.

- Communicating defects to the appropriate person or team.

- Recording updated status of defects – this is always done during more formal review techniques, and is optional for others.

- Evaluating and documenting the review quality criteria and identifying whether or not they have been met.

- Evaluating the review results against the review exit criteria to decide whether the work product is to be rejected for major change or simply updated with minor changes.

- Gathering metrics, such as how much time was spent on the review and how many defects were found, to show notionally how the quality of the reviewed item has increased, but also the value added by the review to later stages of the life cycle; for example, how much the defect would have cost if it hadn't been found until user acceptance testing.

- Checking on exit criteria – the moderator will also check the exit criteria (for more formal review types such as inspections) to ensure they have been met so that the review can be officially closed.

- Accepting the work product when the exit criteria have been met.

ROLES AND RESPONSIBILITIES

The role of each reviewer is to look at the work product(s) under review (and any appropriate source documents) from their assigned perspective; this may include the use of checklists. For example, a checklist based on a particular perspective (such as user, maintainer, tester or operations) may be used, or a more general checklist (such as typical requirements problems) may be used to identify defects.

In addition to these assigned review roles, the review process defines specific roles and responsibilities that should be fulfilled for formal reviews. These are:

- Manager – the manager decides what is to be reviewed (if not already defined), assigns staff, ensures that there is sufficient time allocated in the project plan for all of the required review activities, monitors ongoing cost-effectiveness of the review, determines if the review objectives have been met and executes control decisions if objectives are not met. Managers do not normally get involved in the actual review process unless they can add real value – for example, they have technical knowledge that is key to the review.

- Author – the author is the writer or person with chief responsibility for the development of the work product(s) to be reviewed. The author will in most instances also take responsibility for fixing any agreed defects.

- Moderator (often called a facilitator) – ensures effective running of the review. The moderator may mediate between the various points of view and is often the person on whom the success of the review rests. The moderator will also make the final decision as to whether to release an updated work product.

- Scribe (or recorder) – collates anomalies from reviewers and records review information such as time taken and defects found. The scribe may also attend the review meeting and document all of the issues and defects, problems and open points that were identified during the meeting, which may include new anomalies.

- Reviewers – individuals with a specific technical or business background (also called subject-matter experts, checkers or inspectors) who, after the necessary preparation, identify and describe findings (e.g. defects) in the product under review. As discussed above, reviewers should be chosen to represent different perspectives and roles in the review process and take part in any review meetings.

- Review leader – the review leader is the person who leads the review of the work product(s), including planning the review, running the meeting and follow-ups after the meeting.

An additional role not normally associated with reviews is that of the tester. Testers have a particular role to play in relation to reviews. In their test analysis role, they will be required to analyse work products to enable the development of tests. In analysing the work products, they will also review them; for example, in starting to build end-to-end scenarios they will notice if there is a 'hole' in the requirements that will stop the business functioning, such as a process that is missing or some data that is not available at a given point. Effectively a tester can either be formally invited to review a work product or may do so by default in carrying out the tester's normal test analysis role.

CHECK OF UNDERSTANDING

1. Identify three benefits of reviews.
2. What happens during the planning phase of a review?
3. Who manages the review process?

TYPES OF REVIEW

A single document or related work product may be subject to many different review types: for example, an informal review may be carried out before the document is subjected to a technical review or, depending on the level of risk, a technical review or inspection may take place before a walkthrough with a customer.

Figure 3.2 shows the different levels of formality by review type.

Each type of review has its own defining characteristics. We identify four review types to cover the spectrum of formality. These are usually known as follows.

1. Informal review (least formal) (e.g. buddy check, pairing, peer review)

Key characteristics:

- The main purpose is detecting potential defects (e.g. anomalies).
- A possible additional purpose could be to generate new ideas or to quickly resolve a problem.
- There is no formal process underpinning the review.

Figure 3.2 Formality of reviews

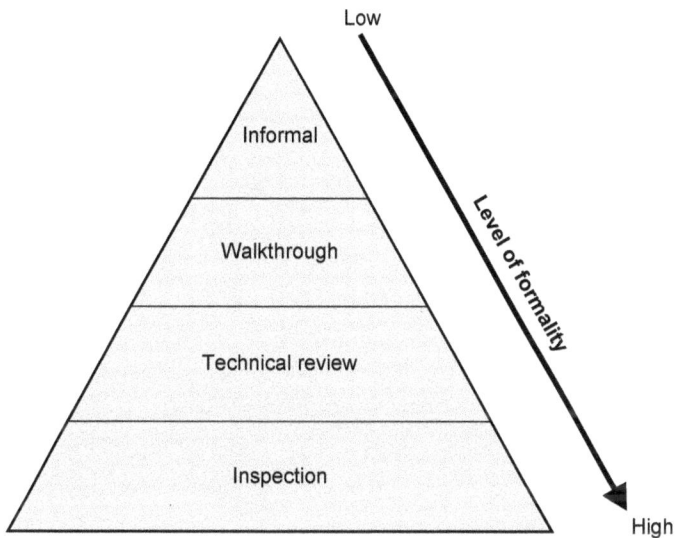

- It may not involve a review meeting.
- It may be performed by a colleague of the author or opened up to many people.
- The review may be documented, but this is not required; many informal reviews are not documented.
- There may be some variations in the usefulness of the review, depending on the reviewer; for example, the reviewer does not have the technical skills but is available to check quickly and ensure that the document makes sense.
- Use of a checklist is optional.
- This is a common review used in Agile development.

2. Walkthrough

Key characteristics:

- The main purpose is to find defects, improve the software product, consider alternative implementations and evaluate conformance to standards and specifications.
- Possible additional purposes include exchanging ideas about techniques or style variations, the training of participants and achieving consensus.
- Preparation by reviewers before the walkthrough meeting, production of a review report or a list of findings, and appointment of a scribe who is not the author are all optional components.
- The meeting is led by the author of the work product under review and attended by members of the author's peer group.

- A scribe is mandatory.
- Use of a checklist is optional.
- Review sessions are open-ended and may vary in practice from quite informal to very formal.
- Walkthroughs typically explore scenarios or conduct dry runs of code or processes.
- Potential defect logs and review reports may be produced.
- It may vary in practice from quite informal to very formal.

3. Technical review

Key characteristics:

- Performed by technically qualified reviewers and led by a moderator.
- The main purpose is gaining consensus and detecting potential defects.
- Possible additional purposes are evaluating quality and building confidence in the work product, generating new ideas, motivating authors to improve future work products and considering alternative implementations.
- Technical reviews are documented and use a well-defined defect detection process that includes peers and technical experts.
- Individual preparation before the review meeting is required.
- A review meeting is optional; if one takes place it is ideally led by a trained facilitator.
- A scribe is mandatory, ideally not the author.
- Reviewers using checklists is optional.
- Potential defect logs and review reports are typically produced.

4. Inspection (most formal)

Key characteristics:

- The main purpose is to detect potential defects, evaluate quality and build confidence in the work product, preventing future similar defects through author learning and root cause analysis.
- A possible further purpose includes authors improving future work products and the software development process, achieving consensus.
- The inspection process is formal, it follows all of the stages of the ISO/IEC 20246 review stages. It is based on rules and checklists, and uses entry and exit criteria.
- Pre-meeting preparation is essential, and includes the reading of any source documents to ensure consistency.

- Reviewers are either peers of the author or experts in other disciplines that are relevant to the work product.

- Specific entry and exit criteria are used.

- A scribe is mandatory.

- Review meetings are led by a trained facilitator who is not the author, and usually involve peer examination of a document. Individual inspectors work within defined roles.

- The author cannot lead the review or be a scribe.

- All potential defects are logged and a review report is produced.

- Metrics are collected and used to improve the entire software development process, including the inspection process.

In reality, the lines between the review types often get blurred, and what is seen as a technical review in one company may be seen as an inspection in another. The above is the 'classic view' of reviews. The key for each company is to agree the objectives and benefits of the reviews that they plan to carry out.

A single work product during its development may be subject to many different types of review.

APPLYING REVIEW TECHNIQUES (NOT EXAMINABLE)

As with testing, there are specific techniques that can be used with all of the aforementioned types of review. The effectiveness of the techniques will vary, depending on what type of review they are used on. Examples of review techniques are as follows.

Ad hoc

Used mainly in the less formal review types, in an ad hoc review those involved are provided with little or no guidance on what is expected of them and how the review should be performed. This technique is very dependent on the reviewer's skills and leads to issues such as duplication of potential defects identified.

Checklist-based

Uses a checklist to deliver a systematic approach to a review, ensuring that the reviewers detect potential defects based on the checklist rules. A review checklist documents the activities to be undertaken, or the types of defects to be identified (e.g. typos or differences to source documents or technical content).

Scenarios and dry runs

A reviewer of a scenario and dry runs is given structured guidelines on how to read the document under review. The scenario allows dry runs of products to take place based on expected usage of the work product. A scenario provides better guidelines on how to identify specific defect types and is sometimes seen as better than the checklist approach.

Role-based

In a role-based review, the work product is examined from a specific role perspective. Typical roles include end user administrator and system administrator, among others.

Perspective-based

A perspective-based review is a little like a role-based review in that the reviewer will take on a different stakeholder's viewpoint. Perspectives such as end user, marketing, design, testing or operations may be required. This approach leads to more depth in the review as well as a reduction in the duplication of potential defects found.

In addition, this type of review requires the reviewer to attempt to deliver the result of the work product. For example, a tester may have to produce draft acceptance tests if reviewing a requirement specification.

This approach has been shown to be the most effective general review technique.

Success factors for reviews

A successful review will depend on using the right type of review and techniques. There are other factors that can influence the successful outcome of a review. These fall into two categories:

1. organisational;
2. people related.

Examples of organisational success factors are:

- Each review should have a clearly predefined and agreed objective and the right people should be involved to ensure that the measurable outcome is met. For example, in an inspection, if the objective is technically based, each reviewer will have a defined role and have the experience to complete the technical review; this should include testers as valued reviewers. Any defects found are welcomed and expressed objectively.
- Review types (both formal and informal) that are suitable to the type and level of work products and reviewers (this is especially important in inspections).
- Review techniques such as checklist-based or role-based reviewing provide effective defect identification of the work product being reviewed.
- Checklists are used to ensure focus on areas of main risk and are up to date.
- Reviewing large documents in small chunks, providing frequent feedback on defects as early as possible to the author.
- Review participants have sufficient time to prepare, which may include undertaking any training required, as well as reading all supporting work products.

- Reviews are scheduled with adequate notice.
- Management support is essential for a good review process (e.g. by incorporating adequate time for review activities in project schedules).
- Facilitation of all review meetings

Examples of people-related success factors are:

- The right people are involved to ensure that the objectives are met; for example, people with different skill sets from the relevant user community.
- The use of testers as valued reviewers so that they can learn about the work product, enabling the development of better and more effective tests, and to be able to develop those tests early.
- Adequate time is allocated to each participant.
- Chunking of the work product into smaller sections enables the reviewer to focus and not lose attention during the actual review meeting.
- Defects found should be acknowledged, appreciated as helping the author and handled objectively.
- The review meeting is well managed, so that it is seen as a valuable use of time.
- There is an atmosphere of trust during the review; there is no evaluation of the person, only the work product.
- The use of negative body language can show boredom, exasperation or hostility and should be avoided.
- Ensuring that adequate training is provided, especially for the formal review processes such as inspections.
- The culture of review is all about learning and process improvement, and is promoted.

CHECK OF UNDERSTANDING

1. Compare the differences between a walkthrough and an inspection.
2. Name three characteristics of a walkthrough.
3. Identify at least five success factors for a review.

SUMMARY

In this chapter we have looked at how review techniques and static analysis fit within the test process defined in **Chapter 1**. We have understood that a review is a static test – that it is a test carried out without executing any code (by reading and commenting on any work product such as a requirement specification, a piece of code or a test plan/test case). We have also looked at the different types of review techniques available, such as walkthroughs and inspections, as well as spending time understanding the benefits of reviews themselves.

Reviews vary in formality. The formality governs the amount of structure and documentation that surround the review itself.

To obtain the most benefit from reviews, they should be carried out as early in the project life cycle as possible, preferably as soon as the document to be reviewed has been written and definitely, in the case of work products such as requirement specifications and designs, before any code is written or executed. The roles of the participant reviewers need to be defined and, in the more structured review techniques, written output from reviews is expected.

We have learned that static analysis is checking the developed software code before it is executed, checking for defects such as unreachable (dead) code and the misuse of development standards. We have also learned that static analysis is best carried out using tools, which are referenced in **Chapter 6**.

Example examination questions with answers

E1. K1 question
Which of the following is most likely to be examined using static testing?

      A.   User guides.
      B.   Defect reports.
      C.   Test logs.
      D.   Attendance reports.

E2. K2 question
Which of the following has the typical formal review activities in the correct sequence?

      A.   Kick-off, review initiation, review meeting, planning, follow-up.
      B.   Kick-off, planning, review meeting, issue communications and analysis, rework.
      C.   Planning, review initiation, individual review, communications and analysis, fixing and reporting.
      D.   Planning, individual preparation, review initiation, individual review, follow-up, fixing and reporting.

E3. K2 question
Which of the following statements are true?

    i.   Defects are likely to be found earlier in the development process by using reviews.

    ii.   Walkthroughs require code but static analysis does not require code.

    iii.   Informal reviews are used to detect potential defects.

    iv.   Ad hoc techniques need lots of preparation time.

    v.   Dynamic testing occurs after reviews have been used to find defects.

    A.   i, ii, iv.
    B.   ii, iii, v.
    C.   i, iv, v.
    D.   i, iii, v.

E4. K2 question
Which of the following defects could be identified by static testing?

    A.   Execution defects, coding defects and security vulnerabilities.
    B.   Coding defects, requirements defects and security vulnerabilities.
    C.   Security vulnerabilities, test basis issues and environment defects.
    D.   Design defects, user defects and test basis issues.

E5. K1 question
Which *one* of the following roles is typically used in a review?

    A.   Champion.
    B.   Author.
    C.   Project sponsor.
    D.   Custodian.

E6. K2 question
Which of the following is a success factor for reviews?

    A.   The total count of lines of code.
    B.   Walkthrough of a requirements document.
    C.   Large documents reviewed as a whole.
    D.   Each review has clear objectives.

Answers to the self-assessment questions in the chapter

SA1. The correct answer is A.

SA2. The correct answer is C.

SA3. The correct answer is D.

Answers to example examination questions

E1. The correct answer is A.

The other answers could be examined in a static review; only A is identified in the syllabus.

E2. The correct answer is C.

The correct sequence is: planning, review initiation, individual review, communications and analysis, fixing and reporting. All of the other options have either the activities in the wrong order or activities missing from the strict flow.

E3. The correct answer is D.

The other answers are incorrect because:

   ii. Walkthroughs do not require code and static analysis does require code.

   iv. Ad hoc reviews need little preparation.

E4. The correct answer is B.

All other options have a dynamic testing defect among the options.

E5. The correct answer is B.

The author is the only role that is typically used in a review. A champion might sponsor the review process but is not a defined role within an actual review; a project sponsor, if technically competent, might be asked to play a defined role within the review process, but while in that role they will not be a project sponsor; finally, a custodian might ensure that the results are stored safely but is not involved in the review itself.

E6. The correct answer is D.

Each review has clear objectives. The remaining answers are not success factors for reviews.

4 TEST ANALYSIS AND DESIGN

John Kurowski

INTRODUCTION

This chapter addresses the area of test analysis and design. We begin by looking at the test development process to give an understanding of where you create test conditions, test cases and test procedures, and then explore the application of test techniques that support the creation of test conditions and test cases. The final section discusses collaboration-based test approaches for defect detection.

Learning objectives

The learning objectives for this chapter are listed below. You can confirm that you have achieved these by using the self-assessment questions immediately following the learning objectives, the `Check of understanding' boxes distributed throughout the text and the example examination questions provided at the end of the chapter. The chapter summary will remind you of the key ideas.

Each learning objective is allocated a K number to represent the level of understanding required; see the **Introduction** (pp. 2–3) for an explanation of K numbers.

Test techniques overview

- FL-4.1.1 (K2) Distinguish black box, white box and experience-based test techniques.

Black box test techniques

- FL-4.2.1 (K3) Use equivalence partitioning to derive test cases.
- FL-4.2.2 (K3) Use boundary value analysis to derive test cases.
- FL-4.2.3 (K3) Use decision table testing to derive test cases.
- FL-4.2.4 (K3) Use state transition testing to derive test cases.

White box test techniques

- FL-4.3.1 (K2) Explain statement testing.
- FL-4.3.2 (K2) Explain branch testing.
- FL-4.3.3 (K2) Explain the value of white box testing.

Experience-based test techniques

- FL-4.4.1 (K2) Explain error guessing.
- FL-4.4.2 (K2) Explain exploratory testing.
- FL-4.4.3 (K2) Explain checklist-based testing.

Collaboration-based test approaches

- FL-4.5.1 (K2) Explain how to write user stories in collaboration with developers and business representatives.
- FL-4.5.2 (K2) Classify the different options for writing acceptance criteria.
- FL-4.5.3 (K3) Use acceptance test-driven development (ATDD) to derive test cases.

Self-assessment questions

The following questions are to enable you to determine your current level of understanding of the topics in this chapter. You will find the answers at the end of the chapter.

Question SA1 (K2)
Match the test work product to the phase in which it is created within the fundamental test process.

- a. test analysis.
- b. test design.
- c. test implementation.

- 1. test case.
- 2. test condition.
- 3. test procedure.

- A. a1, b2, c3.
- B. a2, b1, c3.
- C. a3, b2, c1.
- D. a3, b1, c2.

Question SA2 (K2)
Which of the following statements about statement and branch coverage is true?

- A. 100 per cent branch coverage guarantees 100 per cent statement coverage.
- B. 100 per cent statement coverage guarantees 100 per cent branch coverage.
- C. 100 per cent branch coverage guarantees 75 per cent statement coverage.
- D. As branch coverage covers branches and statement coverage covers statements, there is no link between the two types of white box coverage.

Question SA3 (K2)
Which of the following is not part of the usual format of a user story?

    A. As a [ROLE].
    B. Because of [CONSTRAINT].
    C. I want to [GOAL].
    D. So that [TARGET].

THE TEST DEVELOPMENT PROCESS

We will begin by looking at the process we need to go through to generate the various elements of a test.

As introduced in **Chapter 1**, test creation takes place during the test analysis, test design and test implementation stages of the test process. In doing this, we create **test conditions**, **test cases** and **test procedures**. The following explains why tests are broken down into these three elements.

For this section, we will use an example. **Figure 4.1** shows a simple age entry screen for a caravan holiday website. Our mission is to test that it is working correctly.

Figure 4.1 The age entry screen

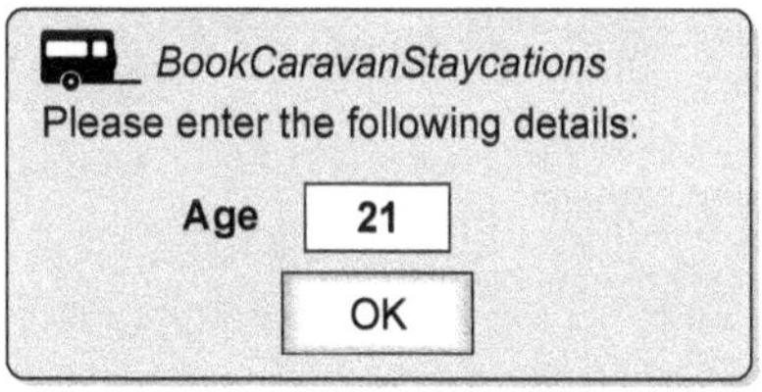

The screen has the following entry criterion to move to the next screen you must enter an age of 18 years old or over.

Test analysis

The first step is **test analysis**. To do this, we need to find something to analyse to use as inspiration for our tests and a source of expected results. This will be the **test basis**, which can be described as any project document you use to base your tests upon.

It is often the case that the quality of the test basis is not sufficient for our needs; it may be poorly written, much too high level, very out of date or may not even exist, which will make determining expected results problematic, to put it mildly. In this case, ideally you would locate a subject-matter expert (SME) and ask them for the information you need or, failing that, revisit the current or legacy system to see what it does. The test basis, SMEs and legacy/current systems are all known as **test oracles**, which are sources that help us to determine the expected results of a test.

For the purposes of this example, our test basis will be requirements, which can be written in many ways. If it was a more traditional or sequential project, then the requirement may appear as:

The system must only allow entry to customers aged 18 years or over.

If it was an Agile project, or iterative incremental, then it could be a user story:

As a customer who wants to book a holiday
And I am 18 years old or over
I want to gain entry into the booking screen
So that I can book a caravan

Either way, we need to analyse the test basis to generate test conditions, as shown in **Figure 4.2**.

Figure 4.2 Test analysis

The ISTQB Glossary describes a test condition as 'A testable aspect of a component or system identified as a basis for testing.' Another way to look at it is as a short statement describing what you want to test, which could be written in the format 'To show that ... or the 'Given ... when ... then ... format.

Taking the first version of the requirement:

The system must only allow entry to customers aged 18 years or over.

The test condition could be written as:

To show that the system must only allow entry to customers aged 18 years or over.

Or could be written in the given–when–then format:

> **Given** I am on the home page
> **And** I am 18 years or older
> **When** I enter my age
> **And** press OK
> **Then** I will be given access to the booking screen.

Either way, we have identified an aspect of the system or component that we can test. We now need to identify the data or information we need for this condition, which is achieved during **test design**.

If we were performing unscripted testing (e.g. exploratory – see the later section) then this may be all the testing we need to specify for test execution, performed by experienced testers.

Test design

The next step is **test design** (**Figure 4.3**). Here we need to consider the data and information we need to run the test condition(s). The ISTQB Glossary definition of a test case is 'A set of preconditions, inputs, actions (where applicable), expected results and postconditions, developed based on test conditions.' From this definition we can extract the main pieces of information we need to fill out the test case:

- **Preconditions**: details or states that need to be set up before we try to execute the test case. For example, if we were testing logging in with a valid user ID and password, we would need these set up in the system database first.

- **Inputs, actions**: information we type in or actions we need to perform during the test.

- **Expected results**: what the system or component should do after processing the inputs.

- **Postconditions**: what else happens after we have run the test case. For example, if we are testing a valid login, there's a very good chance our login will be recorded on some sort of database audit record.

Figure 4.3 Test design

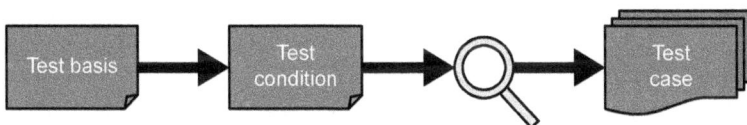

99

When using the given–when–then format, the expected results and postconditions are the **then**, the inputs and actions are the **when**, and the preconditions the **given**.

For the age entry example, **Figure 4.4** shows some possible test cases created using **equivalence partitioning** (much more on that later).

Figure 4.4 Age entry test cases

Preconditions	Inputs	Expected results	Postconditions
On home page	21 years old	Accept	Displays main page
On home page	16 years old	Reject	Stay on home page

If you were using behaviour-driven development (BDD) and following the given–when–then format, then the test cases may look like the examples shown in **Figure 4.5**.

Figure 4.5 Given–when–then examples

Scenario: Old enough	Scenario outline: Booking screen
Given I am on the home page **And** I am **"21"** years old. **When** I try to enter my age **Then** the booking screen will **"accept"**. Scenario: Too young **Given** I am on the home page **And** I am **"16"** years old. **When** I try to enter my age **Then** the booking screen will **"reject"**.	**Given** I am on the home page **And** I am **"<years>"** years old. **When** I try to enter my age **Then** the booking screen will **"<response>"**. Examples: | years | response | | 21 | accept | | 16 | reject |

Test implementation

The next step, having derived the test conditions and test cases, is to generate the **test procedure**, as shown in **Figure 4.6**. This will contain the steps or the script that we need to run during test execution. The ISTQB Glossary definition of a test procedure is 'A sequence of test cases in execution order, and any associated actions that may be required to set up the initial preconditions and any wrap up activities post execution.' Note that a test procedure could cover more than one (related) test case.

Figure 4.6 Test implementation

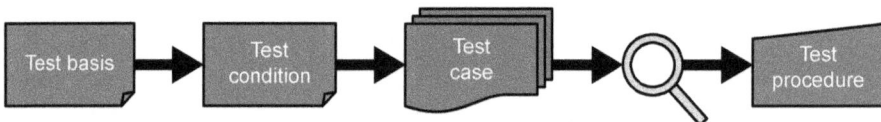

Determining the level of detail to place in the steps really does depend on the context of the testing itself. If we have experienced testers, they will not need as many steps as they should know what they are doing.

A test procedure for a booking example could be as follows:

Enter your age into the home page and click OK – check expected results.

or:

Navigate to the home page
Enter your age in the age field
Click OK
Check expected results.

or (if you were using automation):

```
automationCode script = new automationCode();
@Given("^I am on the home page$")
public void I_am_on_the_home_page() throws Throwable {
  script.openHomePage();
}
@And("^I am \"([^\"]*)\" years old$")
public void I_am_years_old(String arg1) throws Throwable {
  script.checkAge(arg1);
}
@When("^I try to enter my age$")
public void I_try_to_enter_my_age() throws Throwable {
  script.clickOK();
}
@Then("^the website will \"([^\"]*)\".$")
public void the_booking_screen_will(String arg1) throws Throwable
{
  script.assertResponse(arg1);
}
```

Traceability

The last aspect of the test development process is traceability, as shown in **Figure 4.7**. The idea is that if we give everything unique identifiers then they are all traceable backwards and forwards to each other. This helps with keeping the test artefacts (e.g. test cases) up to date should the requirements change. It is also useful for reporting the current status as well as providing information on test basis coverage not only during test execution but also during the test specification steps.

Figure 4.7 Traceability

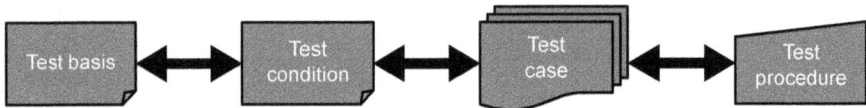

CHECK OF UNDERSTANDING

1. In what order would we create test cases, test conditions and test procedures?

2. In which test work product would we define the expected results?

3. Why is traceability such a great idea?

TEST TECHNIQUES OVERVIEW

The first question we need to ask is: Why use test techniques when designing our test conditions, test cases and test data?

If we go back to the testing principle that exhaustive testing is impossible, we realise that we often don't have enough time or budget to do all the testing that we would like to do. Using test techniques, we can apply a model to the system and that model will suggest a finite number of tests. Not only does this ultimately produce a much smaller number of tests that need to be run, but it can also help to provide coverage statistics.

If we are all taught and apply the test techniques in the same way, then we will have a standardised, consistent approach for designing tests that are easier to review and provide the most efficient coverage, thereby reducing costs.

It can also be quicker. Rather than taking a lot of time to derive these tests from first principles, we can rapidly apply a technique and speed up test creation.

There are three categories of test techniques, as shown in **Figure 4.8**.

Black box test techniques

Black box techniques are also known as **specification-based techniques** and involve treating the system as a black box. The key point to note is that you cannot see inside the box.

Figure 4.8 Categories of test technique

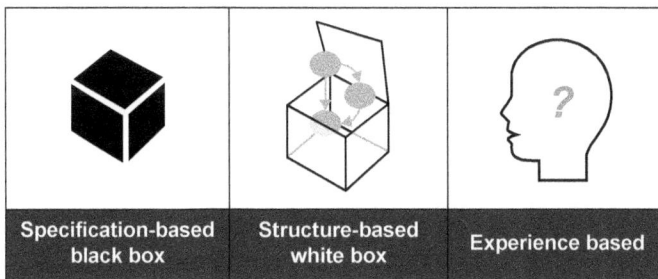

| Specification-based black box | Structure-based white box | Experience based |

For example, a DVD player could be considered a black box. When a DVD is inserted and **play** is pressed, the film appears. We don't know how it does it, just that when certain inputs are performed, certain outputs will result. Now, if the remote had a button labelled, say, **inventory**, we could just press it to see what happens, or, better still, read the manual (i.e. the 'specification').

Another way to describe black box techniques is as 'behavioural', as we are testing to see that the system behaves the way we are expecting it to when we give it certain inputs. The test cases are independent of the software implementation and so, if the software changes, the test cases may not need updating if the behaviour is the same.

Referring back to test types, black box test techniques enable us to test the system functionality (i.e. what the system does) as well as some of the non-functional aspects (i.e. how the system does it) including, for example, load and performance.

White box test techniques

White box used to be known as glass box or clear box. The main point is that you can see inside the box. When you look inside the box, you can see how it is built as you are looking at the structure, which is why these techniques are also known as **structure-based**. White box techniques are purely looking at coverage: How much of the structure have we covered?

For example, if we needed to structurally test a swivel chair, we would look at the way it has been built. If it was on castors, we would include a test where we move it about on the floor. We would include a test to make sure the seat swivels around. If it had two levers, one for the seat to move up and down and the other to adjust the back angle, then we would include tests to check they work. We are looking at the way the chair is built and ensuring that we have a test to check each part of its structure.

At the component level, structure-based testing would usually represent code coverage. If we have a requirement to provide a certain level of code coverage, it is easier to achieve this earlier during component testing rather than during later test levels.

White box test techniques only measure coverage and are applied after the software design is delivered. Also, they do not show whether the system is fit for purpose. Black box test techniques address this aspect, as we will see.

Experience-based test techniques

Experience-based test techniques rely upon having experienced testers who can create tests using their own testing knowledge, experience and natural intuition for finding bugs. They are supplementary 'what have we missed?' techniques, ideally performed after the creation of black box and white box tests.

Another use for experience-based test techniques is when there is not enough time to write the tests before the code arrives, meaning we effectively need to write and run the test during test execution. This will be covered when we talk about exploratory testing.

CHECK OF UNDERSTANDING

1. What would you use to derive test cases using black box test techniques?

2. Which test technique category does not test that the system is fit for purpose?

3. Which test technique category is supplementary, with test cases derived after completion of the other technique categories?

BLACK BOX TEST TECHNIQUES

We will now discuss different black box test techniques.

The techniques we will be looking at are:

- equivalence partitioning;
- boundary value analysis;
- decision table testing;
- state transition testing.

Equivalence partitioning

Equivalence partitioning (EP) is a test technique that organises data into distinct groups known as partitions or equivalence partitions. The idea is that, by selecting one test case in that partition, you assume that the application treats any other value in the same partition the same way. If that one test case finds a defect, then you should find the same defect with any other value in that same partition.

$$\text{Coverage} = \frac{\text{number of partitions tested}}{\text{number of partitions identified}} \times 100 \text{ per cent}$$

The easiest way to demonstrate this will be with a few examples.

EXAMPLE 1: ROOM-BOOKING WEBSITE

The caravan website, as shown in **Figure 4.9**, enables you to state how many rooms you need. The website will only allow you to enter a number in the range of 1 to 10 rooms.

Figure 4.9 The number of rooms example

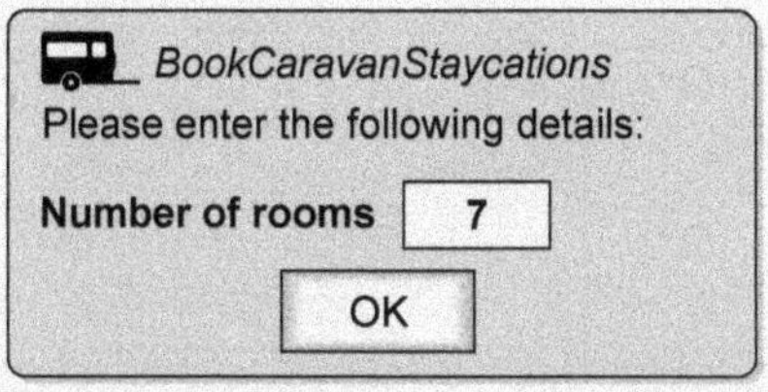

The first question to consider is: How many groups of numbers do we need to worry about? The answer to this is 3. One group covers the invalid numbers below 1, a second group will cover the valid numbers from 1 to 10, and a third group will cover the invalid numbers above 10. There are many other groups we could consider at this point – for example, groups of letters, groups of numbers with decimal places – but for now we will concentrate on whole numbers. One way to display this is as a number line, as per **Figure 4.10**.

Figure 4.10 A number line

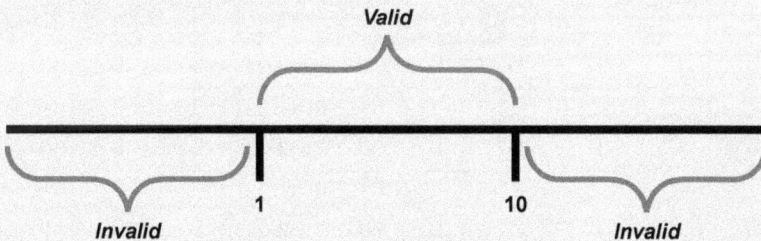

The valid section of the number line covers the input range from 1 to 10 (i.e. the valid partition). It is important to be aware of where one partition ends and the next one begins. Hence there is a section of the number line representing an invalid partition for values less than 1 and another section representing an invalid partition for numbers greater than 10. In EP an invalid partition is one in which the values should be ignored or rejected by the system. The number line not only shows the inputs, but also the expected results (e.g. Invalid). As well as showing these partitions as a number line, we can also express the partitions in a table, as shown in **Figure 4.11**.

The next question to consider is: If pressed for time, what would be the least number of test cases we would need to design to check that this website is working correctly? Once again, you can see from the number line and the table that the answer is 3.

Figure 4.11 Partitions expressed as a table

Partition	<=0	1–10	>=11
Classification	Invalid	Valid	Invalid
Expected result	Reject	Accept	Reject

Where would we get these test cases from? We need to make sure that each test case represents a different partition, as shown in **Figure 4.12**.

Figure 4.12 The three test cases

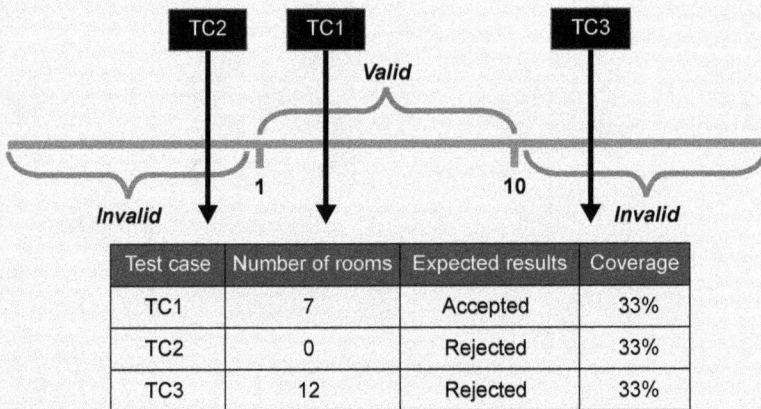

Test case	Number of rooms	Expected results	Coverage
TC1	7	Accepted	33%
TC2	0	Rejected	33%
TC3	12	Rejected	33%

You will see in the test cases in **Figure 4.12** that there is another column showing coverage. As each test case is covering one partition out of three, it is achieving 33 per cent coverage. If you ran TC1 and TC2 then you would cover two out of three partitions, with a coverage of 67 per cent. All three test cases run together would cover three out of three partitions, which would be 100 per cent. You can have many tests cases (e.g. 5, 6, 7, 8 and 9), but if they are all in the same partition then they will not increase the coverage as EP treats each value in a partition the same as any other value in that partition.

EXAMPLE 2: CARAVAN TYPE SCREEN

The website also lets you select the type of caravan you want and its location within the caravan park, as shown in **Figure 4.13**. This time we cannot show the partitions as a number line, but only as lists of options. As the options are radio buttons, the application has already selected the first option in each list. As we cannot unselect all radio buttons from the screen, the options displayed are all valid and there are no invalid partitions.

Figure 4.13 Caravan selection

This time, as we are dealing with options rather than numbers, we can express the various inputs as boxes to select, as shown in **Figure 4.14**. In this partition model, each classification, (i.e. Caravan type and Locations) consists of partitions (or classes) representing just one value.

Figure 4.14 Caravan type selection partitions

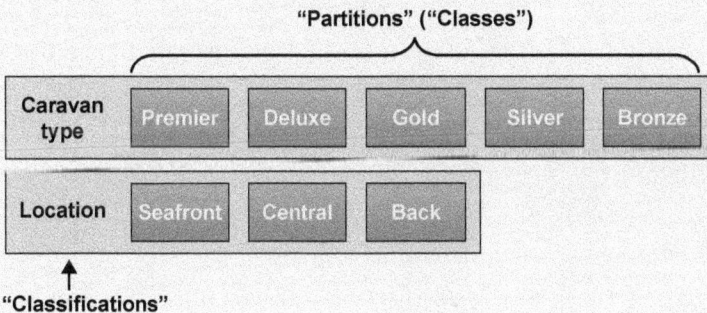

The next question is: How many test cases would we need to cover all partitions? A test case would provide inputs that cover both partitions (e.g. Premier at the Seafront). Initially looking at this, you would think we would need to cover every combination of these partitions, which would be 5 × 3 = 15 test cases, but this could become exhaustive if there were many more classifications. The easiest approach is to ensure you have tested each partition at least once. You may need to cover a partition more than once to achieve this, but this would enable you to cover other untouched partitions. This is what is known as 'each choice coverage' and, as shown in **Figure 4.15**, we can cover all partitions in five test cases.

Figure 4.15 Each choice coverage

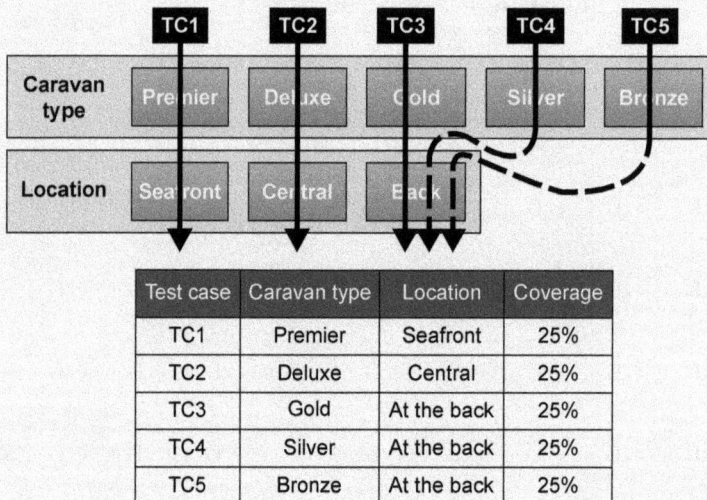

Test case	Caravan type	Location	Coverage
TC1	Premier	Seafront	25%
TC2	Deluxe	Central	25%
TC3	Gold	At the back	25%
TC4	Silver	At the back	25%
TC5	Bronze	At the back	25%

Test cases TC1 to TC3 are covering the first three caravan types and all location choices. TC4 and TC5 are selecting uncovered caravan types, and the second half of those tests could select any location as they have already been covered by the first three test cases. In terms of coverage, each test case is covering two out of a total of eight partitions, which is where we get the 25 per cent figure from.

A quick way to determine the number of test cases needed for 100 per cent coverage will be to look at the classification with the largest number of partitions. In this case it is 'Caravan Type', with five. Therefore, 100 per cent coverage is achievable with five test cases.

We will now look at an example with more than one classification, as well as valid and invalid partitions.

EXAMPLE 3: AN UPDATED ROOM-BOOKING WEBSITE

The website is now updated, as shown in **Figure 4.16**, so that not only do you specify a number of rooms in the range 1–10, but also an age of 18 years or over.

Figure 4.16 An updated room-booking website

This is simply two numbers that could each be represented on a line, although it may be better to show these graphically, as in **Figure 4.17**. The valid partitions have been identified by drawing thick black lines around the boxes.

Figure 4.17 The updated rooms and age partitions

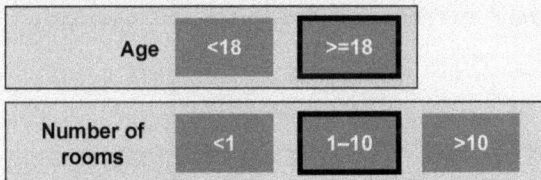

A single test will now select one partition from each classification.

When it comes to identifying the number of tests, your immediate thought is that you could cover these partitions with just three tests (one positive test and two negative tests, as shown in **Figure 4.18**), but it is not that simple.

Figure 4.18 Some possible test cases

Test case	Age	Num rooms	Expected result	Coverage
TC1	21	6	Accepted	40%
TC2	10	0	Rejected	40%
TC3	21	21	Rejected	40%

To determine that each invalid partition is the definite reason why the test produces a negative result, we need to test each one separately. Test case TC2 has been greyed out as we cannot determine which invalid value caused the expected result. It could be the age or the number of rooms. As this is black box testing we are not looking inside the box, so we don't know.

We should not cover two invalid partitions in one test because if one of the partitions was not functioning correctly (i.e. it had a defect), we would not be able to detect this due to the defect in the other invalid partition. Designing a separate test for each invalid partition prevents what is known as error masking. Therefore, we now have four tests, as shown in **Figure 4.19**, with test cases TC1, TC2 and TC4 each representing an invalid partition in the model.

Figure 4.19 The four test cases needed to provide coverage

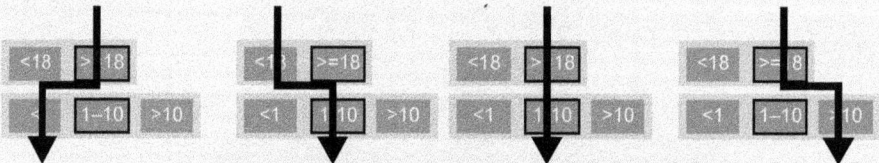

Test case	Age	Num rooms	Expected result	Coverage
TC1	21	0	Rejected	40%
TC2	12	5	Rejected	40%
TC3	21	5	Accepted	40%
TC4	21	21	Rejected	40%

EXERCISE 4.1

Suppose you have a bank account that offers variable interest rates: 0.5 per cent for the first £1,000 credit; 1 per cent for the next £1,000; 1.5 per cent for anything above £2000. If you wanted to check that the bank was handling your account correctly, what valid input partitions might you use?

The answer can be found at the end of the chapter.

Output partitions

So far we have looked at combinations of inputs. We may also need to consider the output partitions as well. Considering this is a black box test technique and we don't know what is happening inside the box, all we can go on are the inputs into the box and the outputs coming out of the box (see **Figure 4.20**). Revisiting Example 3 (see **Figure 4.17**) where we were testing combinations of age and number of rooms, we would need to also consider what the expected result would be when testing these combinations. We could just test the output partitions, in which case with this example we would only need to set up one test for the 'reject' output partition and one for the 'accept' output partition (see **Figure 4.21**).

Figure 4.20 Input and output partitions

Figure 4.21 The updated rooms and age example with input and output partitions

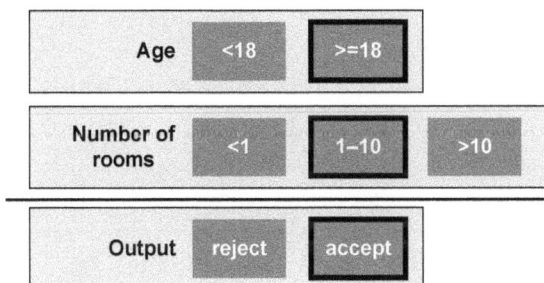

EXERCISE 4.2

A mail-order company selling flower seeds charges £3.95 for postage and packing on all orders up to £20 value and £4.95 for orders above £20 value and up to £40 value. For orders above £40 value there is no charge for postage and packing.

If you were using EP to prepare test cases for the postage and packing charges, what valid partitions would you define?

What about non-valid partitions?

The answer can be found at the end of the chapter.

CHECK OF UNDERSTANDING

1. What is a partition?

2. Why do we only have one test case in each partition?

3. Why perform EP when most people are tempted to test the boundaries instead?

Boundary value analysis

Boundary value analysis (BVA) is a test technique where you reduce the number of test cases you need to run by analysing the values at the boundaries. As a technique, it follows EP but can only work on sorted or ordered partitions (e.g. numbers or alphabetic A–Z).

$$\text{Coverage} = \frac{\text{the number of boundary values that were exercised}}{\text{the total number of identified boundary values}} \times 100 \text{ per cent}$$

Why do we test the boundaries? Because the edges of the ranges are deemed to be higher risk than the middle. In other words, we are more likely to find defects at the point where one range ends and the other begins. Developers are more likely to make mistakes around the handling of these edges – in other words, the boundary values.

Two-value and three-value

There are different levels of coverage when identifying BVA tests. In **Figure 4.22** there is an example of a boundary where the value is a whole number. You may want to perform three-value BVA, which contains a value on the boundary, one increment to the left and one increment to the right. Alternatively, you may only want to perform two-value BVA, which considers the value of the boundary and its nearest invalid value. (There is also four-value, where you do another value either side of the two-value – but we won't go there.)

Figure 4.22 Two-value and three-value BVA

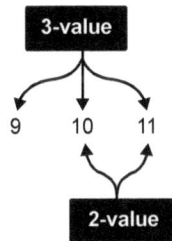

This is the point where you might ask what should you use in the real world? Would you rather use two-value or three-value BVA? There is no right answer to this (you can answer most testing questions with 'it depends').

Two-value BVA would be quicker to apply if, say, you had 1,000 boundaries to test, as you would need 2,000 test cases as opposed to 3,000 for three-value BVA. Although it is quicker, it does make it riskier. Three-value BVA is slower to apply as it requires more tests, but it does give a greater level of confidence.

Figure 4.23 lists the various ways a developer may code the boundary. It should be coded as 'less than or equal to 10' and three-value BVA will find all occurrences of any boundary defect, in this case testing the values of 9, 10 and 11. Two-value BVA in this example would only test the values 10 and 11 and would miss the defect where 9 would be rejected by the incorrect code.

Figure 4.23 Two-value and three-value BVA comparison

2-value		3-value
☑	≤10	☑
☑	<10	☑
☒	=10	☑
☑	≥10	☑
☑	>10	☑

Let's have a look at one of the previous examples again.

EXAMPLE 4: THE ORIGINAL ROOM-BOOKING WEBSITE

Going back to the original room-booking website page, as shown in **Figure 4.24**, the website only allows you to enter a number in the range of 1 to 10 rooms inclusive.

Figure 4.24 The original number of rooms example

BookCaravanStaycations
Please enter the following details:

Number of rooms   7

OK

In order to perform BVA, we need to make sure we have first completed EP so that we can determine the ordered partitions that are present and identify the values at the boundaries of those partitions as per **Figure 4.25**.

Figure 4.25 Identifying the boundaries

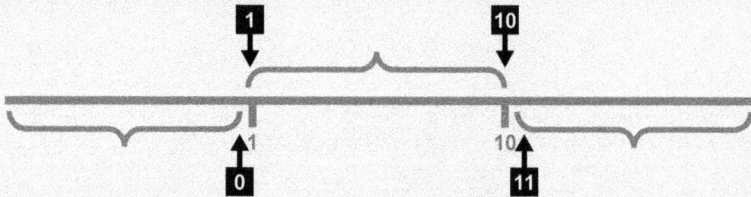

Using two-value BVA

If we were performing two-value BVA on this example, we would be testing four boundary values, as shown in **Figure 4.26**. Each test case would cover one boundary value out of a total of four, which gives a coverage of ¼ or 25 per cent.

Figure 4.26 Two-value BVA

Test case	Age	Coverage
1	0	25%
2	1	25%
3	10	25%
4	11	25%

114

Using three-value BVA

Performing three-value BVA on this example would test six boundary values, as shown in **Figure 4.27**. Each test case will cover one boundary value out of a total of six, which gives a coverage of $1/6$ or 16.67 per cent.

Figure 4.27 Three-value BVA

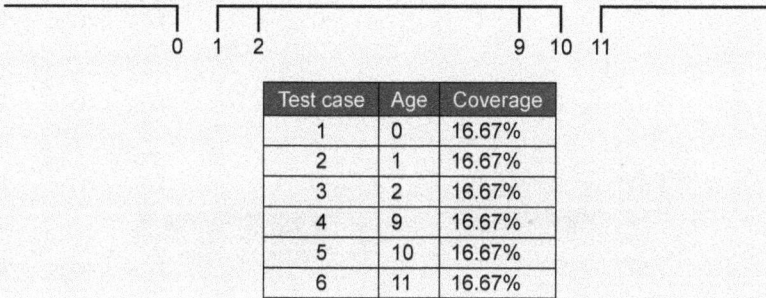

Test case	Age	Coverage
1	0	16.67%
2	1	16.67%
3	2	16.67%
4	9	16.67%
5	10	16.67%
6	11	16.67%

EXERCISE 4.3

A system is designed to accept scores from independent markers who have marked the same examination script, consisting of five questions. Every question is marked out of 20, giving a total for the script of 100. Two markers' scores are compared and differences greater than 3 for any question score or a difference of 10 for the script total are flagged for further examination.

Using EP and BVA, identify the boundary values that you would explore for this scenario.

(In practice, some of the boundary values might actually be in other equivalence partitions, and we do not need to test them twice, so the total number of boundary values requiring testing could be less than you might expect.) The answer can be found at the end of the chapter.

CHECK OF UNDERSTANDING

1. Why do we test the values at the boundaries?

2. Which would you do first, EP or BVA?

3. Why would we perform three-value BVA if two-value BVA is quicker and uses fewer tests?

Decision table testing

So far, with EP and BVA, we tend to look at just testing one value, whereas now we will look at testing combinations of values that provide different outputs. We can do this using decision tables testing.

Decision tables are a very good way of covering complex business rules by analysing the logical conditions between inputs and their expected outputs. They are an effective way of presenting the business logic in a different format that could highlight missing combinations or defects in the requirements. We can use the rules created via these tables as our test cases. As shown in **Figure 4.28**, decision tables consist of four quadrants: a condition stub, a condition entry, an action stub and an action entry.

Figure 4.28 Decision table structure

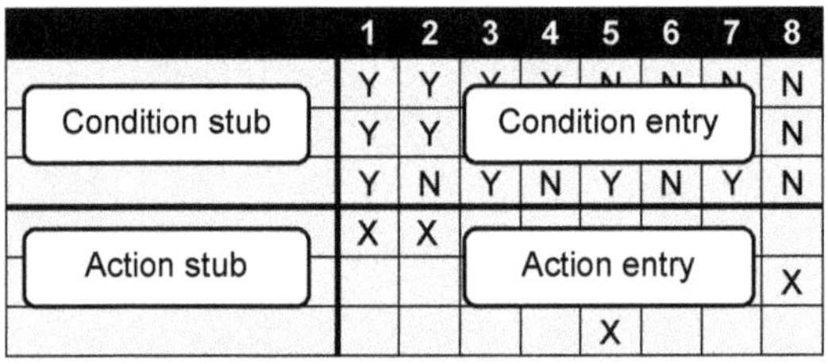

The condition stub is where we list the conditions to test, usually worded in a way that asks a binary question that produces a simple Boolean yes/no answer. With an extended entry decision table (more on that later) the conditions may be expressed more like EP classifications.

The condition entry is where we identify the business rules to test by deriving every possible combination of **true** and **false** for each entry in the condition stub, usually shown as 'T' and 'F' or 'Y' and 'N'. For this kind of decision table, the number of rules work out as two to the power of the number of conditions. For example, with three conditions this is $2^3 = 8$ rules. Each column in the table is a rule and each rule ultimately becomes a test.

The action stub contains a list of all the possible actions that are a result of the combinations of conditions. One rule can have many actions and the number of conditions and actions do not need to match.

The action entry identifies the actions selected by each defined rule. To avoid confusion with the condition entries, the action entries often use a different notation, usually depicted with an 'X' for true and left blank for false.

EXAMPLE 5: AN UPDATED AGE ENTRY WEBSITE

The age entry part of the website had an update. To gain access, not only must you be 18 years old or over, but you must not be a smoker (**Figure 4.29**). The first step we need to do is to identify the conditions and actions.

Figure 4.29 An updated age entry website

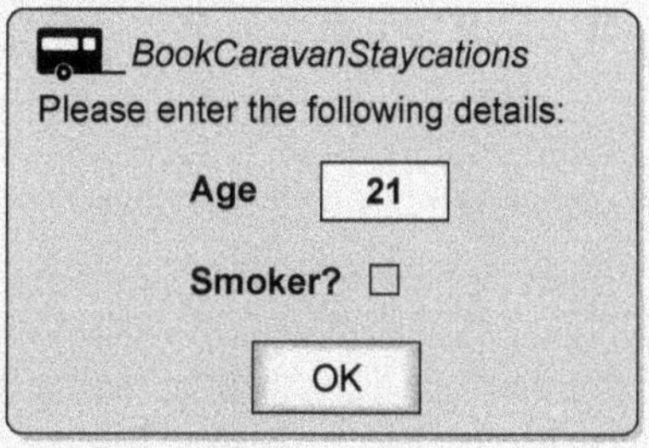

Figure 4.30 shows that we need two conditions to cover the two business rules. It is tempting to define two actions: 'Accept' and 'Reject' (which appeared a lot when we used this example previously for the test development process) but we can cover both outcomes by selecting true or false to the 'Accept' action. We could define both actions, but you would then notice that every time one of them was true, the other action would be false, and vice versa, which adds unnecessary detail to the table. Our next step is to identify all condition combinations and their selected actions.

Figure 4.30 Age entry screen with conditions and actions identified

Figure 4.31 shows the combinations of conditions and actions. Although it is tempting to make the second condition 'Non-smoker', it then becomes confusing when defining the rules as that would become '**True**, they are **not** a non-smoker'. It is easier to define it as: 'Smoker? **False**'. You can also see that we have covered the 'Keep them out' action by selecting false to the 'Accept' action.

Figure 4.31 Age entry screen with rule combinations identified

	1	2	3	4
18 or over?	T	T	F	F
Smoker?	T	F	T	F
Accept		X		

We now have a list of rules that we can take away and execute as our test cases. We can run those four rules to test the system is working correctly. Unlike EP, where we did not design a test case with more than one invalid input value due to error masking, it is acceptable to do so when applying decision table testing as the technique deliberately seeks to test input value combinations.

There can be a situation where you look at a decision table and think: Do we really need to test **all** of these rules? For example, if we have someone who is not over 18, do we really need to test for someone who is a smoker as well as someone else who isn't? The answer to these questions are addressed in the topic of collapsed decision tables.

Collapsed decision tables
One of the issues with decision tables is that, if you are testing many conditions, they can become very big and very complex very quickly. Testing three conditions is achievable as this works out as 8 rules, but 5 conditions is 32 rules and 10 conditions is 1,024 rules. Do we really need to test every possible combination? Is this not, ever so slightly, just a bit exhaustive (remembering from **Chapter 1** that exhaustive testing is impossible)?

The answer is yes. If we revisit the web page there are certain conditions where, when set to 'T' or 'F', it doesn't matter what the other condition is set to as it doesn't change the actions.

There are two ways to collapse this decision table, as shown in **Figure 4.32**. The first way is if they are smokers, they will not be accepted whether they are over 18 or not. The second way is for the age limit, as it does not matter if they are a smoker or not if they are outside this limit. We could compress the table the first way or the second way, but we can't create a table where we compress it both ways (as there would be more than one rule for not over 18 and a smoker). If you notice, rule 1 on the first collapsed table now has a dash in place of the T or F; this indicates that it can be either true or false (or yes or no) – in other words, 'it doesn't matter'. Other options for depicting this are 'Y/N', leaving it blank, or maybe even 'Don't care'.

Coverage
We have looked at generating the decision table, working out if it is collapsible or compressed and how to do this. We have also seen that there can be more than one way to collapse a table. We are now at the point where we take the defined rules and

Figure 4.32 Options for collapsing the decision table

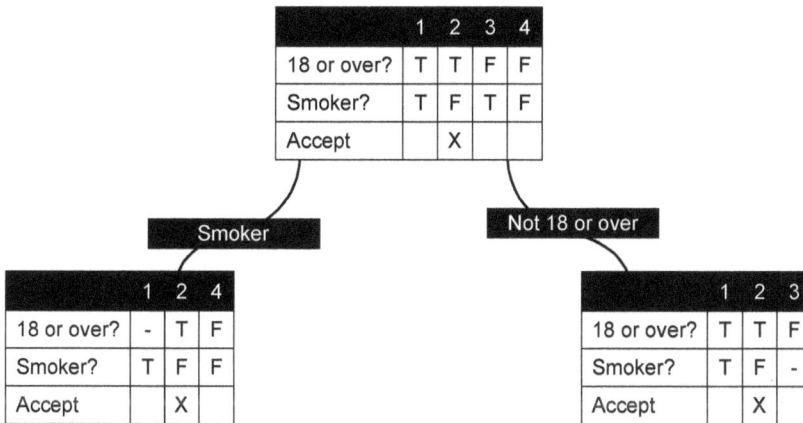

	1	2	3	4
18 or over?	T	T	F	F
Smoker?	T	F	T	F
Accept		X		

Smoker

	1	2	4
18 or over?	-	T	F
Smoker?	T	F	F
Accept		X	

Not 18 or over

	1	2	3
18 or over?	T	T	F
Smoker?	T	F	-
Accept		X	

execute them as test cases. To do this, we will need to report back progress metrics and, to do that, we will need to be able to calculate our coverage.

$$\text{Coverage} = \frac{\text{the number of exercised columns}}{\text{the total number of feasible columns}} \times 100 \text{ per cent}$$

Revisiting the web page example, we will take the original table (known as 'simple rules') and one of the collapsed tables (known as 'complex rules'). On both tables we will run rule number 1, which is somebody under 18 who is a smoker.

As you can see in **Figure 4.33**, collapsing the table or not determines the total number of rules to run and the percentage of coverage each rule would achieve. If we ran rules 1 and 2 on both tables, we would achieve 2 out of 4 = 50 per cent coverage on the simple rules table, and 2 out of 3 = 66 per cent on the complex rules table. Running all rules on a table will give you 100 per cent coverage.

Figure 4.33 Decision table coverage

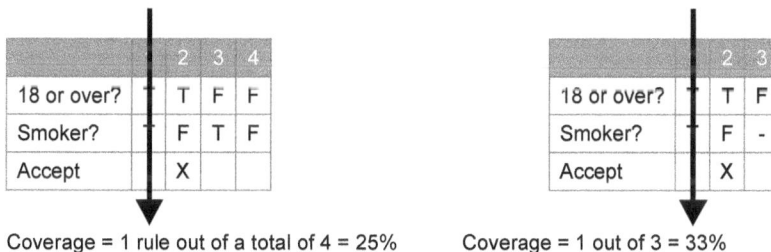

	1	2	3	4
18 or over?	T	T	F	F
Smoker?	T	F	T	F
Accept		X		

Coverage = 1 rule out of a total of 4 = 25%

	1	2	3
18 or over?	T	T	F
Smoker?	T	F	-
Accept		X	

Coverage = 1 out of 3 = 33%

Infeasible combinations of conditions

So far we have looked at an example where all combinations or conditions are possible. There may be times when we need to remove certain combinations of conditions that are unachievable. Let's have another example.

EXAMPLE 6: THE UPDATED ROOM-BOOKING WEBSITE WITH A SENIOR DISCOUNT

We will go back to the updated room-booking website (**Figure 4.34**) where we need to specify a number of rooms in the range 1–10 and an age of 18 years or over. The example has now been updated to include a 10 per cent discount if you are aged 65 or over.

Figure 4.34 An updated rooms website

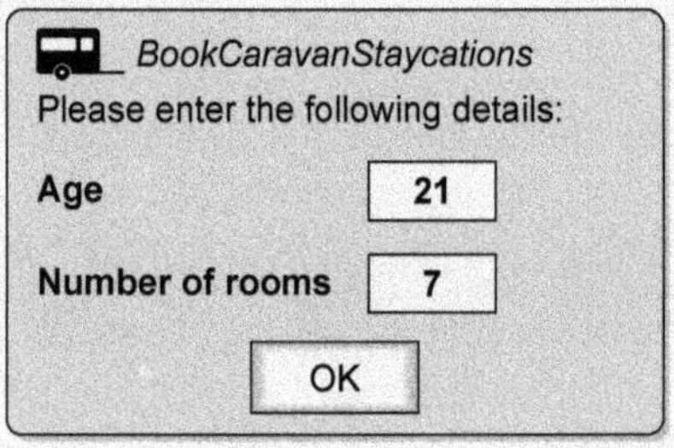

The ranges of values and combinations of conditions to cover are as shown in **Figure 4.35**. We have omitted some of the conditions as we can get to them by setting another condition to **false** – for example, we can select age <18 by setting the condition age ≥ 18 to **false**. We now have three conditions and therefore $2^3 = 8$ rules.

Figure 4.35 A slightly complicated room-booking decision table

	1	2	3	4	5	6	7	8
age >=18	T	T	T	T	F	F	F	F
age >=65	T	T	F	F	T	T	F	F
1 ≥ rooms ≤ 10	T	F	T	F	T	F	T	F
Book caravan	X		X					
Senior 10% discount	X				X	X		

A couple of the rules are infeasible (e.g. you cannot be ≥65 but not ≥18 at the same time). The infeasible rule columns have therefore been greyed out.

Figure 4.36 shows the table with infeasible rules removed, but this decision table is still collapsible. If you look at rules 2, 4 and 8 you will see that if the number of rooms is false, it doesn't matter what the age is set to as the system will not let you book a caravan nor get the discount. The collapsed table is shown in **Figure 4.37**, now with the rules/columns renumbered.

Figure 4.36 A complicated, feasible ticket-booking decision table

	1	2	3	4	7	8
age >=18	T	T	T	T	F	F
age >=65	T	T	F	F	F	F
1 ≥ rooms ≤ 10	T	F	T	F	T	F
Book caravan	X		X			
Senior 10% discount	X					

Figure 4.37 The collapsed, feasible decision table

	1	2	3	4
age >=18	T	-	T	F
age >=65	T	-	F	F
1 ≥ rooms ≤ 10	T	F	T	T
Book caravan	X		X	
Senior 10% discount	X			

Your next question at this point is going to be: Surely there must be an easier way to show this information in a decision table that doesn't take us through compressing so many rules? The answer to this question is with extended entry decision tables.

Extended entry decision tables
So far we have been creating decision tables where the condition rules are set up as true or false. It is also possible to create this table differently, where the condition entries for the rules may contain multiple values (e.g. equivalence partitions, ranges of numbers, specific values).

EXAMPLE 7: CARAVAN DRINKS DISPENSER

The caravan site will have a new machine to provide 24-hour complementary drinks to customers. When a customer first arrives at the caravan site, they will register and be given an ID card. The ID cards fall into three categories: less than 13 years old ('INF'), between 13 and 17 inclusive ('TEEN') and 18 or above ('ADULT').

When a customer wants to select a drink, they insert their ID card and select the appropriate drink. There is a selection between soft drinks ('SFT'), hot beverages ('BEV') and alcoholic drinks ('ALC').

If we were to create a standard decision table, there would be three conditions for the customer type and three for the drink selection. This could generate a table with 64 rules (2 to the power 6). As this would become very confusing, another approach would be to use an extended entry table.

Figure 4.38 shows this example as an extended entry decision table. The actions labelled 'N/A' are infeasible rules. For example, you cannot select an alcoholic drink if you are an infant. **Figure 4.39** shows the table after removing the infeasible rules.

Figure 4.38 An extended entry decision table

	1	2	3	4	5	6	7	8	9
Customer	INF	INF	INF	TEEN	TEEN	TEEN	ADULT	ADULT	ADULT
Drink	SFT	BEV	ALC	SFT	BEV	ALC	SFT	BEV	ALC
Lemonade	X	N/A	N/A	X		N/A	X		
Cola	X	N/A	N/A	X		N/A	X		
Tea		N/A	N/A		X	N/A		X	
Coffee		N/A	N/A		X	N/A		X	
Beer		N/A	N/A			N/A			X
Whisky		N/A	N/A			N/A			X

Figure 4.39 An extended entry decision table with infeasible rules removed

	1	2	3	4	5	6
Customer	INF	TEEN	TEEN	ADULT	ADULT	ADULT
Drink	SFT	SFT	BEV	SFT	BEV	ALC
Lemonade	X	X		X		
Cola	X	X		X		
Tea			X		X	
Coffee			X		X	
Beer						X
Whisky						X

After removing the infeasible rules, it is now possible to see that the decision table is collapsible. As everyone can select soft drinks, we can collapse rules 1, 2 and 4 as shown in **Figure 4.40**.

Figure 4.40 A collapsed, extended entry decision table with infeasible rules removed

	1	2	3	4
Customer	-	TEEN	ADULT	ADULT
Drink	SFT	BEV	BEV	ALC
Lemonade	X			
Cola	X			
Tea		X	X	
Coffee		X	X	
Beer				X
Whisky				X

EXERCISE 4.4

A mutual insurance company has decided to float its shares on the stock exchange and is offering its members rewards for their past custom at the time of flotation. Anyone with a current policy will benefit provided it is a 'with-profits' policy and they have held it since 2001. Those who meet these criteria can opt for either a cash payment or an allocation of shares in the new company; those who have held a qualifying policy for less than the required time will be eligible for a cash payment but not for shares. Here is a decision table reflecting those rules.

	Rule 1	Rule 2	Rule 3	Rule 4
Current policy holder	Y	Y	Y	N
Policy holder since 2001	N	Y	N	-
'With-profits' policy	Y	Y	N	-
Eligible for cash payment	X	X		
Eligible for share allocations		X		

What result would you expect to get for the following test case?

Joyce is a current policy holder who has held a 'with-profits' policy since 2003. The answer is at the end of the chapter.

CHECK OF UNDERSTANDING

1. What is the difference between a simple decision table and an extended entry decision table?

2. Why do we remove infeasible rules from a decision table?

3. What is the point of collapsing a decision table?

State transition testing

So far, with EP, BVA and decision table testing we have looked at techniques that are data driven. We will now look at a technique that is event driven.

State transition testing is looking at the system in a totally different way. In applying this technique we identify the various states that the system can remain in, theoretically indefinitely, until a trigger moves the system to another state (i.e. a transition). It is easier to explain with an example.

EXAMPLE 8: AN ELECTRIC TOOTHBRUSH

We have a two-speed electric toothbrush operated by a single button. Pressing the button once will switch the toothbrush onto the slow speed. A second button press will turn the toothbrush to the fast speed. A third button press will switch it off.

State transition diagram
One way to show this information is graphically, as a state transition diagram (**Figure 4.41**).

Figure 4.41 An example state transition diagram

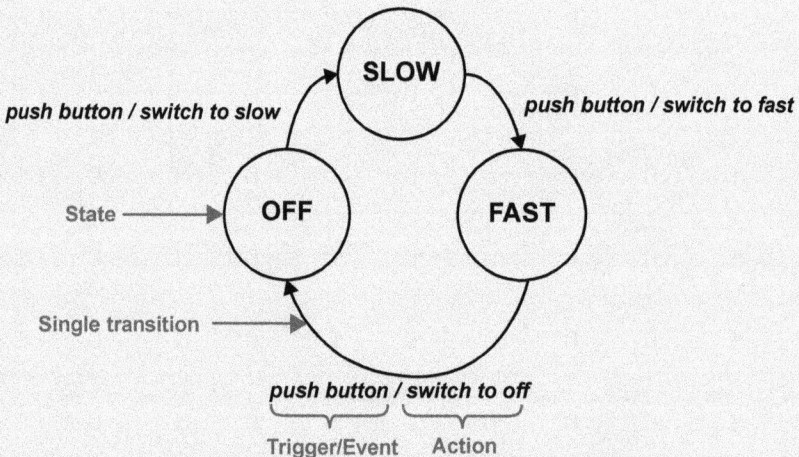

The objects in circles (can be squares) are **states**. This is how the system will remain, theoretically indefinitely, until a trigger or event will move it from this state to another state.

The lines between the states are valid **single transitions**. A single transition moves the system from one state to another. Each transition has two pieces of information:

- The first piece of information before the slash is the **trigger** or **event** that moves the system from one state to another and instigates the single transition. For example, on the diagram we move from the off state to the slow state by pushing the button.

- The second piece of information, after the slash, is the **action**. This shows what will happen upon activating a trigger.

Something to note is that the transition is always labelled 'trigger/event' then 'action', whichever direction the transition is going.

Something else to note is that each transition has one arrow as it will move from one state to another one in only one direction. State transition diagrams should not show a double-headed arrow – instead it should be two separate transitions (as the action or trigger would be different).

State table
Another way to present this information is in a **state table**. In this format we list the different states in the left-hand column and the different triggers or events along the top row. The central part of the table will show the next state reached by actioning the trigger/event from each state. This is shown in **Figure 4.42**. For example, if the toothbrush is currently in the slow state, the push button trigger/event would cause it to transition to the fast state.

Figure 4.42 An example state table

State/Event	Push button
OFF	SLOW
SLOW	FAST
FAST	OFF

Test cases
From here we can generate a test case to cover each transition, as shown in **Figure 4.43**.

Figure 4.43 Some example test cases covering each transition

	TC1	TC2	TC3
Start state	OFF	SLOW	FAST
Trigger	Push button	Push button	Push button
Action	Switch to SLOW	Switch to FAST	Switch to OFF
End state	SLOW	FAST	OFF

It is common practice in state transition testing to create test cases that cover several transitions. Looking at this example, we might ask whether it would be possible to navigate through the diagram and cover all the transitions with just one test case. The answer is yes, and can be expressed as: off–slow–fast–off. It can also be expressed as slow–fast–off–slow or fast–off–slow–fast. The point is we are covering each transition and each state and, as far as the technique goes, it doesn't matter where we start as long as we cover everything.[1]

Let's look at a more complex example.

EXAMPLE 9: AN AUDIO PLAYER

There is a need to test a basic audio player with only Play, Stop, Next and Previous buttons. The equipment plays music when you press the Play button and will stop playing music when you press the Stop button. The Next button will move to the next track and the Previous button will move to the previous track. The Next and Previous buttons only work when the player is playing music. **Figure 4.44** shows a state transition diagram for this example, with all the valid transitions detailed.

---

1   Although, in reality, you would start with the toothbrush in the off position.

Figure 4.44 A state transition diagram for the audio player example

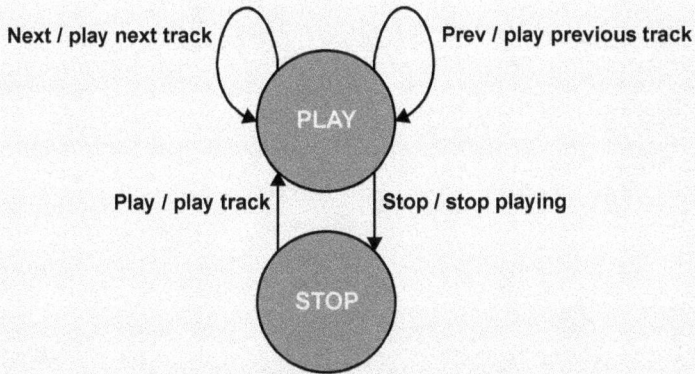

Figure 4.45 shows the same information in a state table. There are several cells with a dash in them, which show that this is an **invalid transition**. For example, if you try to press Play when in the Play state, nothing should happen. It is tempting to fill the cell with the name of the starting state, but this suggests there is a transition that points back at itself. Transitions that point back at themselves, as we see in the diagram, are valid transitions and sometimes referred to as 'pig-ears'.

Figure 4.45 A state table for the audio player example

State/Event	STOP	PLAY	NEXT	PREV
PLAY	STOP	-	PLAY	PLAY
STOP	-	PLAY	-	-

The state table is useful as it shows both valid and invalid transactions. By identifying the invalid transitions, you will be able to include negative test cases as well as positive ones.

Test cases
A test case to cover each valid transition is shown in **Figure 4.46**.

Figure 4.46 Some example test cases covering each transition

	TC1	TC2	TC3	TC4
Start state	STOP	PLAY	PLAY	PLAY
Trigger	Play	Stop	Next	Prev
Action	Play track	Stop playing	Play next track	Play previous track
End state	PLAY	STOP	PLAY	PLAY

We can cover all of these transitions with one test, but it does look confusing if you express it via the state changes as you can't tell which transitions were navigated:

STOP – PLAY – PLAY – PLAY – STOP

In this example, it would be easier to describe the way to navigate using the triggers

Play – Next – Prev – Stop

A test case to cover each invalid transition is shown in **Figure 4.47**.

Figure 4.47 Some example test cases covering each invalid transition

	TC5	TC6	TC7	TC8
Start state	PLAY	STOP	STOP	STOP
Trigger	Play	Stop	Next	Prev
Action	Invalid	Invalid	Invalid	Invalid
End state	N/A	N/A	N/A	N/A

Let's look at another example.

EXAMPLE 10: A WEATHER GAUGE

We have a new weather gauge that can display the weather in two ways: as a barometer or as a thermometer. The device has three buttons. The Display button will switch the device between displaying barometer details and thermometer details and vice versa. The Config button will take you into the appropriate configuration mode – when displaying the thermometer this will be 'thermometer config' and when displaying the barometer it will be 'barometer config'. To return from 'thermometer config' mode to displaying the thermometer details, press the Return button; this button will also return you to the barometer display from barometer config mode.

State transition diagram

To draw this example, we need to work out the states and the trigger or events. The states, as mentioned earlier, would be the state the system could stay in, theoretically forever, until a trigger changes it. Looking at this example, we have the following four states:

- display thermometer details (thermometer);
- display barometer details (barometer);
- thermometer config (thermometer config);
- barometer config (barometer config).

We also have the following triggers or events:

- the display button (display);
- the config button (config);
- the return button (return).

Once we have worked out how the states connect according to the triggers, we are able to draw a state transition diagram (**Figure 4.48**).

Figure 4.48 A state transition diagram for the weather gauge

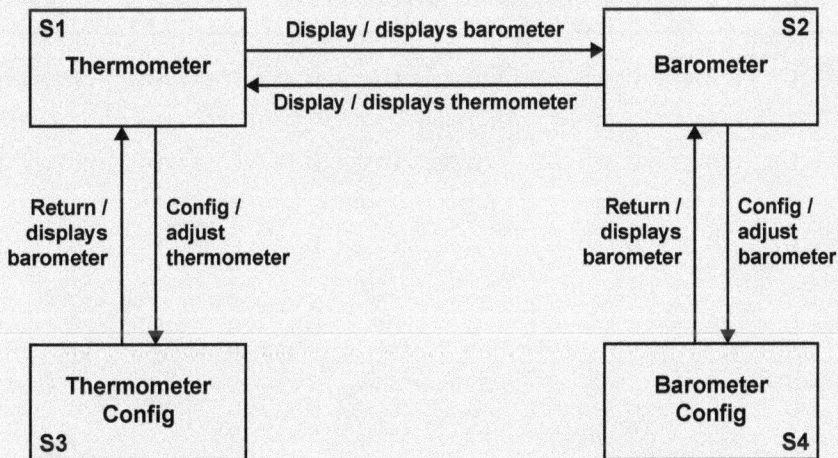

The information in the state transition diagram is also displayable as a state transition table (**Figure 4.49**). The table also shows the invalid transitions, which are not easily visible from the diagram and can be used to design negative test cases. The various states have been labelled S1 to S4 to make the following tables easier to read.

Figure 4.49 A state transition table for the weather gauge

State/Event	Display	Return	Config
S1	S2	-	S3
S2	S1	-	S4
S3	-	S1	-
S4	-	S2	-

Test cases

From here we can generate a test case to cover each transition (**Figure 4.50**).

Figure 4.50 Some example test cases covering each transition

	TC1	TC2	TC3	TC4	TC5	TC6
Start state	S1	S2	S1	S2	S3	S4
Trigger	Display	Display	Config	Config	Return	Return
Action	Displays barometer	Displays thermometer	Adjust thermometer	Adjust barometer	Displays thermometer	Displays barometer
End state	S2	S1	S3	S4	S1	S2

When generating test cases to execute, we look at ways to navigate through the state model.

A test case based on a state transition diagram or state table is usually represented as a sequence of events, which results in a sequence of state changes (and actions, if needed).

One test case may, and usually will, cover several transitions between states. For example, taking **Figure 4.48**, possible test cases could be:

S3–S1–S2–S4–S2–S1–S3

or:

S1–S3–S1–S3–S1–S2–S4–S2–S1–S2

This second test case covers some transitions more than once – for example, S1 to S3 – but this may reflect typical use of the weather gauge and so is worthy of testing.

We now look at guard conditions – let's have another example.

EXAMPLE 11: PRE-BOOKED TRAVEL TICKET MACHINE

At a train station there is a machine that will print tickets that have been pre-booked for travel. The way it works is:

- You enter your booking reference number.
- The machine asks you to insert your payment card.
- After you have inserted your card, the machine scans to see whether you have any tickets to print.
- If you have any tickets to print, it will display a 'Printing ...' message, print the tickets and then return your card.
- If you don't have any tickets to print, it will display a 'No tickets found ...' message and return your card.

Guard conditions
Sometimes, when designing the state transition diagram, there may be a transition that will only occur when a trigger fulfils an attached condition. This transition condition is known as a **guard condition**. They are usually labelled as follows: *'event [guard condition] / action'*.

This example has guard conditions, as shown in **Figure 4.51**. Transitions do not always have to have guard conditions (none of the other examples so far have); they can also be omitted if they are not helpful when testing.

Figure 4.51 A state transition table for the ticket machine

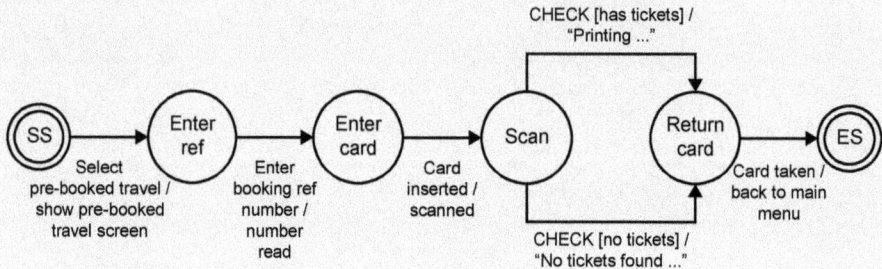

Coverage

We have now looked at several examples of machines and systems that we can test by applying state transition testing. The next question is: How do we provide coverage? The answer is, as with most answers in testing, it depends. There are different levels.

We will use **Figure 4.52**, which is a much simpler version of the weather gauge example (originally **Figure 4.48**) to illustrate the different levels of coverage.

Figure 4.52 A simplified state transition diagram

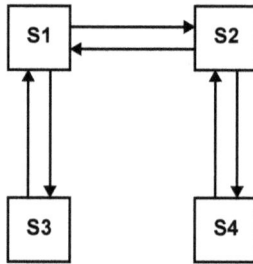

All states coverage 'All states' coverage just covers all states on the diagrams. It does **not** need to cover all of the transitions. We can visit the same state more than once. For this example, we can achieve all states coverage with the following test cases:

$$S3–S1–S2–S4$$

or

$$S1–S3–S1–S2–S4$$

$$\text{Coverage} = \frac{\text{the number of visited states}}{\text{the total number of states}} \times 100 \text{ per cent}$$

Valid transitions coverage 'Valid transitions' coverage covers all valid transitions on the diagram (and therefore also all the states). It does **not** need to cover any invalid transitions. To cover all transitions, we may have to traverse the same transition more than once. For this example, we can achieve valid transitions coverage with the following test cases:

$$S1–S2–S4–S2–S1–S3–S1$$

or

$$S1–S3–S1–S2–S1–S2–S4–S2$$

$$\text{Coverage} = \frac{\text{the number of exercised valid transitions}}{\text{the total number of valid transitions}} \times 100 \text{ per cent}$$

All transitions coverage With 'all transitions' coverage, we need to cover the valid and also try to cover the invalid transitions. To do this, it is easier to look at the state table (**Figure 4.53**) rather than the state diagram.

Figure 4.53 A state transition table for the weather gauge

State/Event	Display	Return	Config
S1	S2	-	S3
S2	S1	-	S4
S3	-	S1	-
S4	-	S2	-

To cover the valid transitions we can use test cases for the 'valid transitions coverage' and will then need to add new test cases for all of the invalids. The following is an example list:

- return from S1;
- return from S2;
- display from S3;
- display from S4;
- config from S3;
- config from S4.

Again, to prevent error masking, we would want to run these invalid transitions as separate test cases.

N-switch coverage Another level of coverage is n-switch coverage. The first thing to clarify is what the 'n' in 'n-switch coverage' represents. It is the number of transitions in a sequence minus one. If we were interested in just covering single transitions, there is only one transition per sequence, so n – 1 becomes 1 – 1 = 0-switch coverage. If we were interested in covering sequences of two transitions, then this would be 1-switch coverage (n – 1 becomes 2 – 1 = 1). Sequences of three would be 2-switch and four would be 3-switch. By way of an example, if wishing to travel from London to Leeds by railway, there is regular direct service and hence no need to change trains. It is a 0-switch journey. However, if wanting to travel from London to Bradford, then changing trains at Leeds is probably necessary as there are very few direct services to this latter destination. It would be a 1-switch journey.

Identifying the 0-switches is just a case of noting all the transitions in the state transition diagram. Identifying 1-switch sequences takes longer – there is no easy way to do this – and involves starting at one state, identifying all the series of two transitions from that state (including back to itself), and then moving onto the next state.

Taking the weather gauge example again, **Figure 4.54** lists all 0-switches and 1-switches on the state transition diagram. It also has a test case S1–S2–S4–S2. You can see that this test case covers 3 of the 0-switches out of a total of 6 and gives us 50 per cent coverage. This test case only covers 2 out of 10 1-switches, which gives us 20 per cent 1-switch coverage.

Figure 4.54 n-switch coverage for the weather gauge

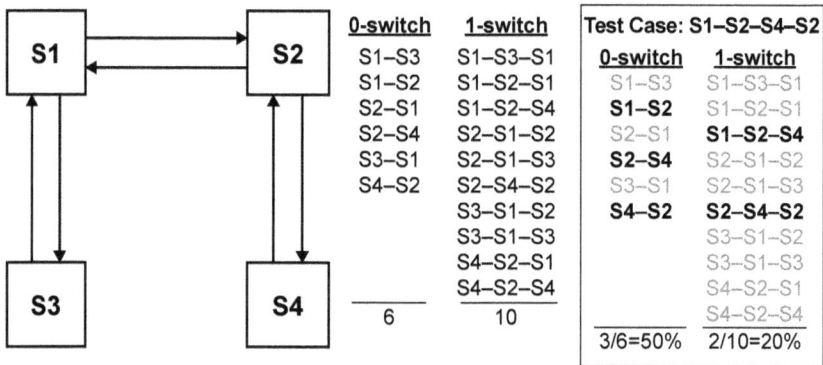

	0-switch	1-switch	Test Case: S1–S2–S4–S2	
			0-switch	1-switch
	S1–S3	S1–S3–S1	S1–S3	S1–S3–S1
	S1–S2	S1–S2–S1	**S1–S2**	S1–S2–S1
	S2–S1	S1–S2–S4	S2–S1	**S1–S2–S4**
	S2–S4	S2–S1–S2	**S2–S4**	S2–S1–S2
	S3–S1	S2–S1–S3	S3–S1	S2–S1–S3
	S4–S2	S2–S4–S2	**S4–S2**	**S2–S4–S2**
		S3–S1–S2		S3–S1–S2
		S3–S1–S3		S3–S1–S3
		S4–S2–S1		S4–S2–S1
		S4–S2–S4		S4–S2–S4
	6	10	3/6=50%	2/10=20%

The big question, then, is which is the better coverage – 0-switch or 1-switch? Well, 0-switch is quicker and coverage easier to achieve, but it could also be riskier; 1-switch coverage would take longer but it would give you greater confidence. There may be defects found using 1-switch that 0-switch would miss. For example, there may be a defect moving from barometer config to displaying the barometer and then straight back into barometer config, which you would definitely find with 1-switch coverage but could be missed with 0-switch coverage.

WHITE BOX TEST TECHNIQUES

Introduction

As explained earlier, a black box test technique will treat the system as a black box. We give it certain inputs and then check that it produces the correct outputs, without being particularly interested in how the system derives the output values. With white box testing, we are interested in what is inside the box. What can we see when we look inside? We see the structure, how it is built. Therefore, white box test techniques are purely based on how much of the structure have we covered.

EXAMPLE 12: A SWIVEL CHAIR

Imagine sitting on a swivel chair – perhaps not exactly like the one shown in **Figure 4.55**, but we'll use this as an example. If we wanted to test this swivel chair structurally, we would inspect the chair and identify all the structural elements to cover with test cases. This chair has wheels, so we would need to test that it can move on them. It swivels,[2] so we'd need to include a swivel test case. There are two levers to cover with test cases – one lever moves the seat up and down and the other moves the backrest forwards and backwards. Although the picture doesn't do it justice, if you pull out the seat lever the seat can spring up to its highest position or down to its lowest (if you are sitting on it), and we'd need to include a test case for that. The other lever can pull out and push in, which holds the back rest in its position. In short, it is a basic swivel chair, and we will need all these test cases to cover it structurally.

---

2   That is why it is called a swivel chair.

Figure 4.55 A swivel chair

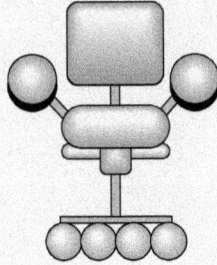

As software testers, we are not usually testing swivel chairs, but will be looking at systems or components. Most of the time we are considering code coverage, but we could easily be looking at menu trees, website structure, business processes, HTML – anything with a structure to it. For this section, we will concentrate on code coverage.

CODE EXAMPLE 1

In **Figure 4.56** we have an example of some code with a simple decision. This is a simple piece of pseudo-code that is not intended to reflect code you would see in the real world, but is here purely to show how the code structure and coverage techniques work. Here the code is going to check that if the TV is on, we will watch it. Next to the code are two ways to draw this code.

Figure 4.56 Code Example 1

```
1. IF TV=ON
2.     WATCH
3. END IF
```

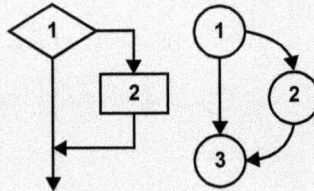

The first way is a normal flow chart; the diamonds are decisions (one line goes in and two lines come out) and the squares are just executable statements (one line in and out). Decisions have two 'decision outcomes', which are **true** and **false**. Here we follow the usual convention of having the line for the true decision outcome (the TV is on) going out to the right and the false decision outcome (the TV is not on) as the line heading straight downwards. You will notice that the 'END IF' (the end of the decision) doesn't have a shape but is on the diagram at the point where the two lines from the two decision outcomes join up to form one line again.

The other way of depicting the code is as a control flow diagram, as shown by the structure comprising just circles. This diagram could be circles or it may be represented as squares. Either way a decision is a shape with two lines coming out of it to cover both decision outcomes, and executable statements have one line both in and out, but you will notice that the 'END IF' is now a shape with two lines going in (technically known as a junction point).

Statement testing and statement coverage
The first level of coverage we need to concern ourselves with is **statement coverage**. To achieve 100 per cent statement coverage, we need to identify the least number of test cases to cover all the shapes on the flow chart. A test will start at the top of the diagram, work its way through the shapes and end at the bottom.

There may be a tendency to say that two test cases are needed in this example, assuming there is a need to cover all of the lines in the flow chart, but that is incorrect. We are just interested in the least number of test cases to cover all the code, which does not require execution of each line in the chart. The false branch, where TV is off, has no code (i.e. shape) underneath it to execute. Looking at **Figure 4.57** and the three lines of code on the left-hand side again, if we make sure TV = on then this one test will execute both shapes. In other words, all lines of executable code. Hence the answer is one, not two test cases.

Figure 4.57 Statement coverage test cases for Code Example 1

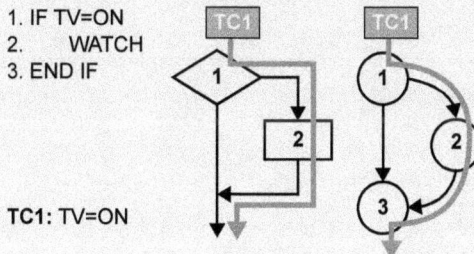

Branch testing and branch coverage

The next level of coverage we need to look at is **branch coverage**. Here, rather like the branches of a tree, we are looking at the various ways through the diagram and trying to cover each way just the once. The easiest way to do this is to concentrate on the decision outcomes: we want to identify the least number of test cases that cover all trues and all falses. A test case only needs to pass through a particular decision outcome (i.e. true or false) once for it to be covered.

Figure 4.58 now shows that we need two test cases: TV = on and TV = off.

Figure 4.58 Branch coverage test cases for Code Example 1

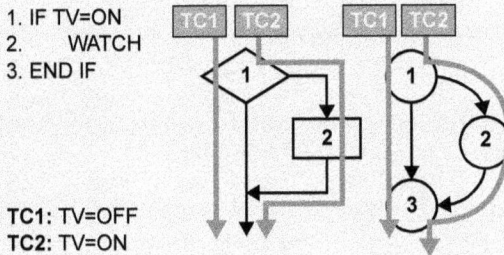

Let's do another one.

CODE EXAMPLE 2

In **Figure 4.59** we now have a simple decision within the code, but this time there is an executable statement down each line of the decision outcome.

Figure 4.59 Code Example 2

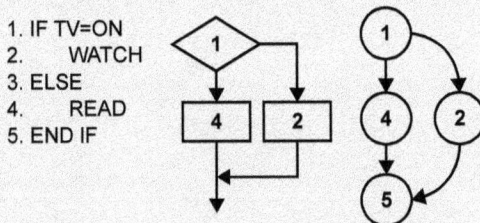

Statement and branch coverage

As you can see in **Figure 4.60**, from a statement coverage point of view we need to go down the true line to cover line 2 and the false line to cover line 4. This is also true for branch coverage. Therefore, for both 100 per cent statement coverage and 100 per cent branch coverage, we need two test cases.

Figure 4.60 Statement and branch coverage for Code Example 2

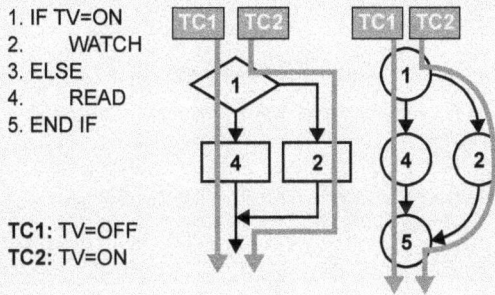

```
1. IF TV=ON
2.    WATCH
3. ELSE
4.    READ
5. END IF


TC1: TV=OFF
TC2: TV=ON
```

Let's have another example, this time with a nested IF.

CODE EXAMPLE 3

A nested IF is where we have an IF statement within another IF statement. **Figure 4.61** shows you the code and the flow chart.[3]

Figure 4.61 Code Example 3: a nested IF

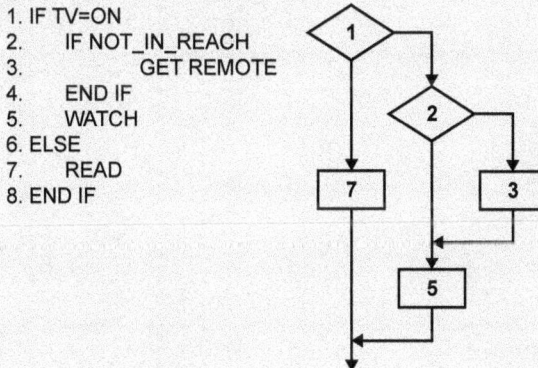

```
1. IF TV=ON
2.    IF NOT_IN_REACH
3.         GET REMOTE
4.    END IF
5.    WATCH
6. ELSE
7.    READ
8. END IF
```

---

3 There is no need to show a control flow diagram for each example, the flow chart will suffice in explaining the technique.

With this example, the extra IF statement revolves around checking whether the remote is within reach. If not, we need to get it before watching the TV. We don't need the remote to read.

Statement coverage

As shown in **Figure 4.62**, this flow chart is slightly deceptive as the NOT_IN_REACH decision does not have an ELSE statement, which means there is no code down the false decision outcome, so we do not need to add a test to go down there to achieve 100 per cent statement coverage.

Figure 4.62 Statement coverage for Code Example 3

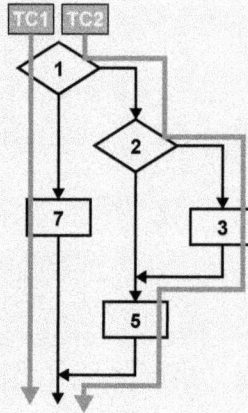

```
1. IF TV=ON
2.     IF NOT_IN_REACH
3.             GET REMOTE
4.     END IF
5.     WATCH
6. ELSE
7.     READ
8. END IF
```

TC1: TV=OFF

TC2: TV=ON
       NOT_IN_REACH=TRUE

Branch coverage

Figure 4.63 shows that we now do need to go down the false decision outcome for the NOT_IN_REACH decision as, although there is no code there, we do still need to cover the false decision outcome.

Figure 4.63 Branch coverage for Code Example 3

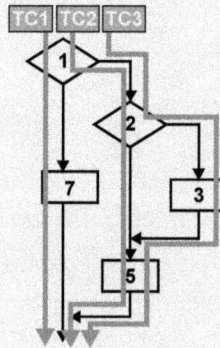

```
1. IF TV=ON
2.     IF NOT_IN_REACH
3.             GET REMOTE
4.     END IF
5.     WATCH
6. ELSE
7.     READ
8. END IF
```

TC1: TV=OFF

TC2: TV=ON
       NOT_IN_REACH=FALSE

TC3: TV=ON
       NOT_IN_REACH=TRUE

In this example, there are three statements, fours decisions and seven branches (where a 'branch' is a line joining two boxes on a structure diagram). We have so far talked about achieving 100 per cent statement coverage or 100 per cent branch/decision coverage. If only test case TC1 is run, this gives differing values for the coverage.

Type	Total	Covered by TC1	Percentage covered
Statement	3	1	33%
Decision	4	1	25%
Branch	7	2	28.56%

CODE EXAMPLE 4

Figure 4.64 is a code example with a loop. This piece of code is addressing how to read a book, including taking it and putting it back into a bookcase. The reading part is covered by the WHILE loop. When running, execution will enter the WHILE loop as long as the decision in the WHILE loop equates to true. As lines 1 and 2 are always executed consecutively, we have grouped them together in the one box. We have done the same with lines 7 and 8.

Figure 4.64 Code Example 4: a loop

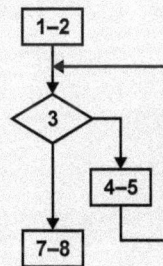

```
1. GET BOOK FROM BOOKCASE
2. OPEN BOOK
3. WHILE NOT AT_END_OF_BOOK
4.    READ PAGE
5.    MOVE TO NEXT PAGE
6. END WHILE
7. CLOSE BOOK
8. PUT BOOK BACK IN BOOKCASE
```

Statement and branch coverage
When you first look at any loop structure you may tend to think that working out all of the test cases to provide 100 per cent code coverage will be incredibly time-consuming and complex, but it isn't. The key thing to note with a loop is that it returns to the original decision again, which enables us to go down the true decision outcome on the first pass through and then, when the WHILE condition changes, be able to automatically traverse the false decision outcome. In short, for any loop we only need one test case for 100 per cent statement or branch coverage, as shown in **Figure 4.65**.

Figure 4.65 Statement and branch coverage for Code Example 4

1. GET BOOK FROM BOOKCASE
2. OPEN BOOK
3. WHILE NOT AT_END_OF_BOOK
4.   READ PAGE
5.   MOVE TO NEXT PAGE
6. END WHILE
7. CLOSE BOOK
8. PUT BOOK BACK IN BOOKCASE

TC1: AT_END_OF_BOOK=FALSE

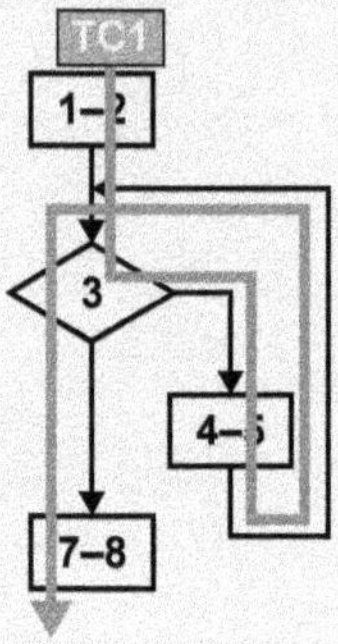

CODE EXAMPLE 5

Figure 4.66 is another code example showing a SELECT CASE statement, which is an easier way to navigate through a list of IF and ELSEIF statements.

Figure 4.66 Code Example 5

1. INPUT "Enter month", MONTH
2. SELECT CASE MONTH
3.   CASE 1 TO 3
4.   SEASON = " is in WINTER"
5.   CASE 4 TO 6
6.   SEASON = " is in SPRING"
7.   CASE 7 TO 9
8.   SEASON = " is in SUMMER"
9.   CASE 10 TO 12
10.   SEASON = " is in AUTUMN"
11.   CASE ELSE
12.   SEASON = " is INVALID"
13. END SELECT
14. PRINT MONTH & SEASON

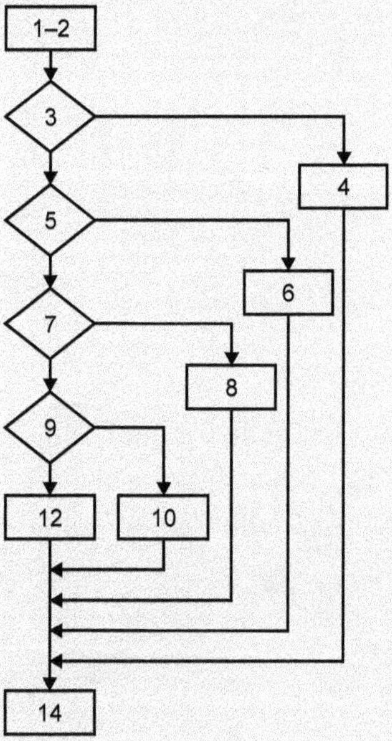

Statement and branch coverage

When looking at coverage for this example, the number for statement coverage and branch coverage will be the same as there is code down each line of every decision diamond in the flow chart, as shown in **Figure 4.67**.

Figure 4.67 Statement and branch coverage for Code Example 5

1. INPUT "Enter month", MONTH
2. SELECT CASE MONTH
3. CASE 1 TO 3
4. SEASON = " is in WINTER"
5. CASE 4 TO 6
6. SEASON = " is in SPRING"
7. CASE 7 TO 9
8. SEASON = " is in Summer"
9. CASE 10 TO 12
10. SEASON = " is in AUTUMN"
11. CASE ELSE
12. SEASON = " is INVALID"
13. END SELECT
14. PRINT MONTH & SEASON

TC1: MONTH = 0
TC2: MONTH = 11
TC3: MONTH = 8
TC4: MONTH = 5
TC5: MONTH = 2

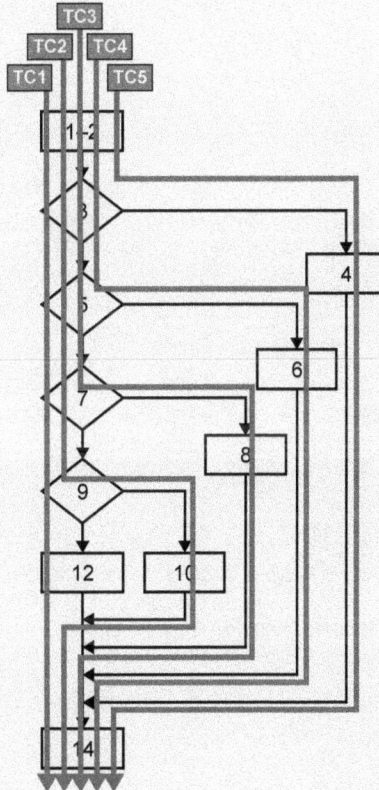

CODE EXAMPLE 6

This example is expressed in simple pseudo-code, which may seem a little scary to begin with. Please don't worry about what the code is trying to achieve. Variables labelled A to F are used in the code. It is the structure we are concerned with. **Figure 4.68** is a decision followed by another decision.

Figure 4.68 Code Example 6

```
1. IF A>B
2.     E=B-A
3. END IF
4. IF C>D
5.     F=D-C
6. END IF
```

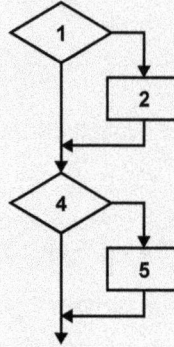

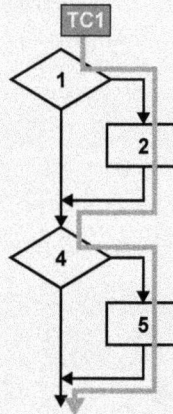

Statement coverage

The next question is: How many test cases do we need for 100 per cent statement coverage? It is a decision followed by another decision. Looking at the code, you need to design the least number of test cases to cover all of the shapes on the flow chart. This can be achieved with just one test, as shown in **Figure 4.69**.

Figure 4.69 Statement coverage for Code Example 6

```
1. IF A>B
2.     E=B-A
3. END IF
4. IF C>D
5.     F=D-C
6. END IF
```

TC1:   A=5, B=1, C=9, D=1

Branch coverage

The next question is: How many test cases do we need for 100 per cent branch coverage? Remembering that to achieve branch coverage we need to concentrate on the decisions and decision outcomes. We need to identify the fewest test cases that cover all of the true and false outcomes.

As you can see in **Figure 4.70**, we can achieve 100 per cent branch coverage with two test cases. The TC1 branch covers the two trues and TC2 covers a branch with both falses. If the first test was true then false, this would still be okay as long as the second test was false then true.

Figure 4.70 Branch coverage for Code Example 6

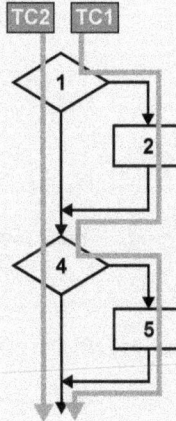

```
1. IF A>B
2.      E=B-A
3. END IF
4. IF C>D
5.      F=D-C
6. END IF
```

TC1:   A=5, B=1, C=9, D=1
TC2:   A=1, B=3, C=1, D=6

It may be tempting to think that we would require four test cases for 100 per cent branch coverage, which as we have seen is incorrect. However, this is the number for 100 per cent path coverage.

Path coverage
Although the Foundation Certificate qualification doesn't specifically address path coverage, it is worth understanding as there is an actual hierarchy of white box test techniques. Path coverage is every possible way through the flow chart and does become very complex very quickly. **Figure 4.71** shows the test cases we need, which is basically every combination of true and false for both decisions. Although this doesn't look too bad for this example, if we copied and pasted that decision again so we had one decision followed by another decision and then another one, we would then need 2 × 2 × 2 = 8 test cases for path coverage, and we are starting to head towards exhaustive testing once again. You may think 8 is not too bad, but if we copied it again it then becomes 16, then 32, then 64 – you get the idea. If we had that same decision flow chart 10 times in a row it would then be 1,024 test cases to achieve 100 per cent path coverage.

Figure 4.71 Path coverage for Code Example 6

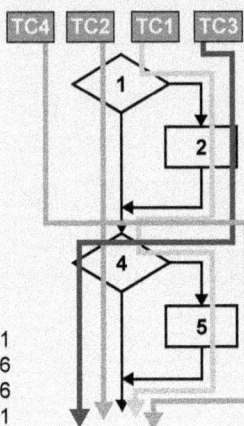

1. IF A>B
2.   E=B-A
3. END IF
4. IF C>D
5.   F=D-C
6. END IF

TC1: A=5, B=1, C=9, D=1
TC2: A=1, B=3, C=1, D=6
TC3: A=5, B=1, C=1, D=6
TC4: A=1, B=3, C=9, D=1

The value of white box testing

In most real-world situations the perceived level of risk would be too low to justify the use of path coverage. Nonetheless, it is a technique worth considering. **Figure 4.72** shows the white box testing hierarchy. If we have 100 per cent path coverage then we will definitely have 100 per cent branch coverage. If we have 100 per cent branch coverage, then we will definitely have 100 per cent statement coverage. It is described as one 'subsuming' the other one. Path coverage subsumes branch coverage and branch coverage subsumes statement coverage. By 'subsumes' we mean 'includes all of the test cases of the lower white box test technique coverage plus possibly some more'. If you revisit the code examples we have used, you will see that the number of test cases for 100 per cent branch coverage is either the same as 100 per cent statement coverage or requires more tests, but you will **never** find a flow chart where you need more test cases for 100 per cent statement coverage than you do for 100 per cent branch coverage. It is not possible.

Your next question will probably be: If 100 per cent branch coverage is so much better than 100 per cent statement coverage in terms of testing code structure, then why bother with statement coverage? The answer is that statement coverage requires fewer test cases than branch coverage and is therefore quicker. That said, although it is quicker, it is riskier. Branch coverage gives you greater confidence than statement coverage and could also find more code defects. For example, if one decision was coded as 'IF A>5 OR A<100', this will always be true as a number less than 5 is also less than 100 and a number greater than 100 is also greater than 5. You could never test the false decision outcome. The decision as to which technique to use is influenced by the factors of risk and cost, like many other decisions in testing.

Figure 4.72 The white box testing hierarchy

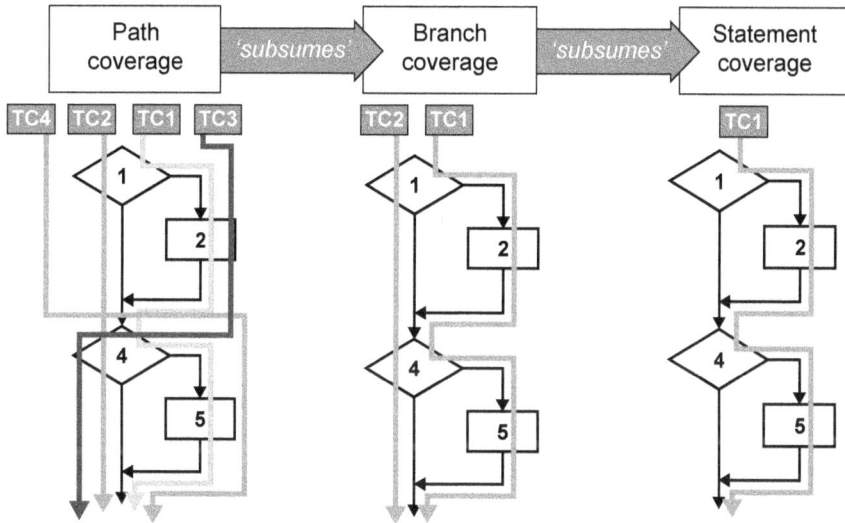

Using white box techniques during static testing

Black box and white box testing are both dynamic testing techniques: we execute the test cases to check that the actual results match the expected results (be it inside the box or outside). White box techniques sometimes get confused with static testing. In the real world, once you have identified the test cases for statement or branch coverage, perhaps as part of the review process, you would then run them (in dynamic testing) to achieve the actual coverage.

That said, when we are looking at the code with a view to identifying test cases for statement or branch coverage, we are performing a static analysis technique known as control flow analysis.[4] This involves reviewing the code without execution to navigate ways to flow through it and hopefully find early defects. For this reason, and this reason only, you **could** say that you can use white box techniques in static testing (e.g. during a dry run of the code).

CHECK OF UNDERSTANDING

1. Which type of coverage concentrates on the number of executable lines of code run?

2. Which is the better coverage: branch coverage or statement coverage?

3. Which coding structure would need more white box tests: IF ... THEN ... ELSE or WHILE ... END WHILE?

---

4   This is covered in the ISTQB Certified Tester Advanced Level Technical Test Analyst syllabus.

EXPERIENCE-BASED TEST TECHNIQUES

We will now look at experience-based test techniques, which rely on having experienced testers. Up until this point we have looked at systematic test techniques that can be applied without testing experience but with the necessary technique training. You can teach people the black box and white box test techniques we discussed in the previous sections, where they can analyse the test basis, generate a finite number of test cases based on one or more techniques, run them and thus provide a specific level of coverage. An experienced tester would also apply these techniques and in doing so probably find defects, but their experience (of current and previous systems) would act as a 'test basis' for the design of additional test cases worth executing. These test cases would be based upon their experience, knowledge, intuition, testing skills and (often the case) paranoia and gut instinct. They would supplement the systematic test cases and can be identified by employing experience-based test techniques.

There can be various reasons why we need to employ experience-based test techniques:

- We may have to use our experienced testing brains where there is no test basis to hand at a suitable level of detail to create 'systematic' test cases.
- The level of risk is considerably higher in certain areas of the system[5] and needs more thorough testing.
- There are certain combinations of values, not covered by the test techniques, which need to be checked.
- The techniques don't cover values that, say, 75 per cent of the business uses.
- Historically we have witnessed a lot of defects in the area under test.
- The developer or business analyst working on that area of the system lacks experience and mistakes are anticipated.

The experience-based test techniques we will be looking at in this section are:

- error guessing;
- exploratory testing;
- checklist-based testing.

These techniques, in essence, enable the experienced tester to intelligently apply test techniques that are black box or white box to the context of the application under test.

Error guessing

Error guessing is based on knowledge of previous defects[6] and will vary from one experienced tester to another, as the ability to identify what might go wrong is heavily dependent on the testing that has been performed so far. Defects found when testing

---

5   Also known as: I'm very worried about those bits.

6   Technically is should be 'defect guessing' as humans make mistakes, and you are testing the manifestations of any mistakes.

computer games will be different to those found when testing heart monitoring equipment and testing the software for an online auction site. Examples of defects found by this technique could include:

- leaving mandatory fields blank;
- filling a number field with as many 9s as possible;
- inputting 29 February when it isn't a leap year;
- finding duplicate key errors in a database by creating, then deleting, then recreating the same new customer;
- entering numbers in alphabetic fields, letters in numeric fields, etc.

The list is endless.

The advantage of employing this test technique is that you can generate test cases that no one else has thought of and find more defects that way. The disadvantage is that it is heavily reliant on your own experience. If you have never tested this kind of application or computer platform before (e.g. you have only tested web browsers on PCs but not on mobile phones) then you would probably miss defects. Employing error guessing with more than one experienced tester helps to address this issue.

A systematic, methodical approach to error guessing is called **fault attacks**. This involves looking at the object to test, generating a list of ways in which this object could fail and then generating test cases to try to make the object fail that way. These test cases are known as 'attacks' as we are looking at various ways to attack the test object. If more than one (experienced) tester is involved in this activity, they could conduct a **fault attack workshop**, where they create a list of possible defects collaboratively.

We will return to error guessing later in this chapter, after checklist-based testing, by way of an example.

Exploratory testing

Exploratory testing is essentially exploring the system under test, testing an area not covered by pre-written test cases, making a note of what you do as you do it, seeing what it does and whether it is okay or whether there are defects to report.

There are a number of reasons why exploratory testing can be of particular benefit:

- exploring how bad a defect is (e.g. the logo disappears on the web page);
- where there is not enough time to write the test cases before the system arrives;
- where there is no test basis or documentation available (or it is of poor quality), which means we need to rely upon our experience to find defects;
- letting an experienced user loose on the system to test usability or to just see what they do (e.g. an end user 'using' the latest release).

It is also usefully employed on development life cycles, such as Agile, where change is embraced and there is a need to be able to adapt quickly and not waste unnecessary time creating test cases ahead of schedule that, should priorities keep changing, become irrelevant.

There are advantages to exploratory testing. An experienced tester is likely to know where the weak spots in the system are and can find the more important defects quicker. They can perform effective exploratory testing when there are time pressures or little test basis documentation available. This technique is also useful for testing system usability to determine how easy it would be for end users to use the system in the way that supports their day-to-day jobs. This is known as 'heuristic evaluation'.

There are also disadvantages: it is very hard to determine coverage: Did we cover all the requirements, and could we have missed vital areas? It is also problematic reproducing defects unless we have made extensive notes as we tested.

A more structured approach to exploratory testing is **session-based testing (Figure 4.73)**. A **session** is an uninterrupted piece of testing, performed within a **time-box** (otherwise we could test all day), and based on a **charter**. The charter can represent one or many test conditions and contains the system area or functionality we wish to test to help satisfy test objectives. For example, we could spend 30 minutes testing the screen navigation on the main menu. The details of the session are usually recorded on a session sheet and, at the end of the session, the tester goes through a debrief where they go through the session test results with a stakeholder or another tester to check they are satisfied with the testing conducted.

Figure 4.73 Session-based testing

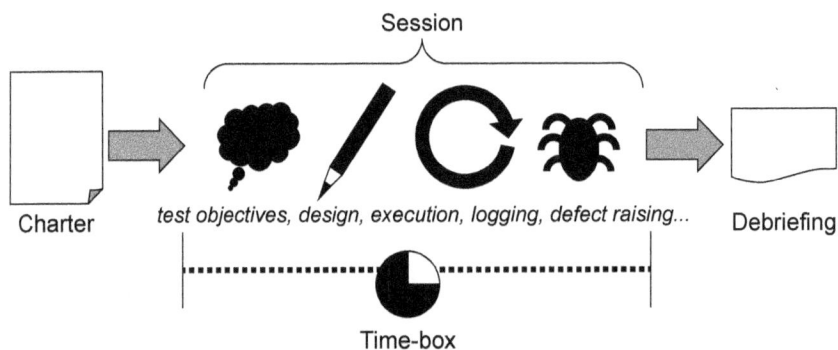

Some people say that exploratory testing is the same as ad hoc testing, but there is one major difference. Exploratory testing is a focused piece of testing. It focuses on a charter and all testing done relates to this charter. If you discover a brand new avenue of testing, this then becomes another session with another charter. Ad hoc testing, on the other hand, is unfocused and tests whatever comes into your head at the time. If everything was ad hoc, then it would be very difficult to work out what testing you had performed previously as the logged results could be located anywhere, if at all. With exploratory testing it would be easier to identify the testing performed as it would fall under a specific session and the results would be more structured and organised.

Checklist-based testing

Checklist-based testing is based on a checklist. There's not a lot more to add. Checklists are usually a list of questions ('Have you thought of this?' 'Have you thought about that?') that an inexperienced tester can work their way through to design and execute test conditions. As an experienced tester you would more likely become involved in designing the checklists. Examples of items that could be on a checklist may include checking that mandatory fields cannot be blank, entering 29 February when it is not a leap year, etc. There can be checklists addressing specific test types, such as usability or security, or perhaps focused on testing graphical user interfaces. Checklists are developed and maintained to satisfy objectives and goals. They can be designed for specific uses and developed at an organisation level to create a standardised testing approach. There are also checklists for specific industries, and some are commercially available.

The advantages of checklists are: they provide a uniform level of assessing quality ('I have gone through the checklist'); they are reviewable and can be revisited to ensure all testing coverage is achieved; and everyone performs the same checks in the same way and so there is a standard approach.

The disadvantage of checklists is that there may be a temptation for testers to only test what is on the checklist – they don't think outside the box (be it a black or a white one). They may think that, because they have gone through the checklist, they have covered everything. If an item to be tested is not on the checklist, then it won't be tested.

EXAMPLE 13: A SIMPLE LOGIN SCREEN

Figure 4.74 shows a simple login screen. Just enter the username and password and click LOGIN and you should log in.

Figure 4.74 A simple login screen

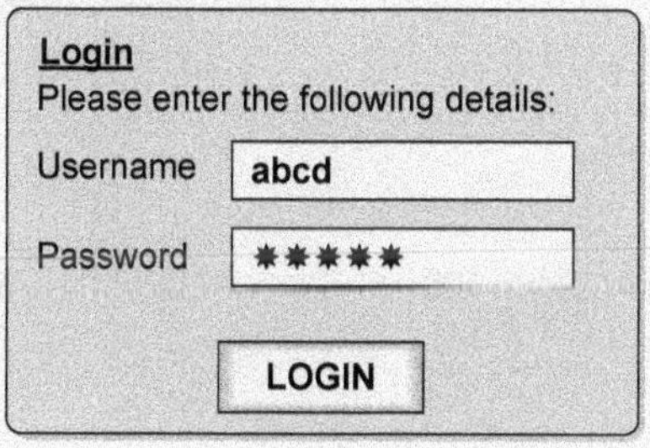

Using either error guessing or exploratory testing there would be many tests that you could perform to test this screen. To begin with, you could leave the fields blank (falls under EP as it is expecting a partition of letters etc.). If the username is up to eight characters long, you could try typing in a nine-character username (falls under BVA). You could do the same with the password if that is also expecting eight characters or more. There are other EP tests for the password – perhaps it requires a combination of upper case, lower case, numeric and special characters ('SnowWhite&the7Dwarfs' is more than eight characters). You could try combinations of valid username and valid password, then valid username and invalid password, etc. (falls under decision table testing). You could see that you stay on the same screen when entering invalid login details and check you move to the main menu when you log in correctly (state transition testing). There is also security to consider and you could try a SQL injection by typing the following into the username field: ADMIN' OR 1;--.[7]

There may also be heuristic-based checklists available (existing within an organisation or found online) that could be used to test the login screen.

SYSTEMATIC AND EXPERIENCE-BASED TECHNIQUES

How do we decide which is the best technique? There are some simple rules of thumb:

1. Always make functional testing the first priority. It may be necessary to test early code products using structural techniques, but we only really learn about the quality of software when we can see what it does.

2. When basic functional testing is complete, that is a good time to think about test coverage. Have you exercised all the functions, all the requirements, all the code? Coverage measures defined at the beginning as exit criteria can now come into play. Where coverage is inadequate, extra tests will be needed.

3. Use structural methods to supplement functional methods where possible. Even if functional coverage is adequate, it will usually be worth checking statement and decision coverage to ensure that enough of the code has been exercised during testing.

4. Once systematic testing is complete, there is an opportunity to use experience-based techniques to ensure that the most important and most error-prone areas of the software have been exercised. In some circumstances, such as poor specifications or time pressure, experience-based testing may be the only viable option.

---

7 On a web page with no HTML validation, if a SQL query is checking login details the 'OR 1' will always return a record and the ';--' comments out the rest of the command.

COLLABORATION-BASED TEST APPROACHES

All the test techniques we have looked at so far are to do with generating test cases to look for defects. This section covers test approaches rather than test techniques and is looking into collaborative ways of preventing or avoiding the defects, rather than detecting them. Any software development life cycle can use these test approaches, but we are looking more into iterative incremental life cycles such as Agile. As it is collaborative, the main expectation is that the whole team will take part.

Collaborative user story writing

User stories can be the equivalent of requirements in a project. They represent a feature desired from the system that provides value to a stakeholder. Usually written from a user role or perspective, user stories tend to be used in Agile software development but could be used in other life cycles.

User stories are most suitable for functional requirements, but can also cover non-functional requirements, although this would probably involve others to help define them at the right level. As they will be one of the main drivers for the project, we need to make sure they are correct and that everyone understands them – hence 'collaborative'.

User stories have three critical aspects, usually called the '3Cs', and are as follows:

- **Card** – the medium that the user story is written down upon. It used to be an actual card but can also be electronic. By making this an actual card, you reduce the amount of information that you can store, making the card hold just enough to understand the need. The idea is that the user story is a placeholder for a further discussion.

- **Conversation** – this is where the team discuss the user story in more depth, gain a shared understanding of what they will need to deliver, and consider the user story from different angles to cover the business, development and testing perspectives.

- **Confirmation** – this is when the team define the acceptance criteria used for evaluation to determine if the goal of the user story has been fulfilled.

The user story is usually in the following format:

> As a [ROLE]
> I want to [GOAL]
> so that [TARGET].

The **role** is the 'who' and is the actor or user affected by the user story. The **goal** is the 'what' and represents the goal they wish to achieve. The **target** is the 'why' and details the resulting business value for the role of the story.

The team write the user stories using various collaborative techniques including brainstorming and mind mapping.

- **Brainstorming** – when the team get together, shout out ideas and one of them writes these up on a flipchart. It is a very creative technique as team members can bounce ideas off each other: one person suggesting something may spark a new thought in someone else and they can then add to it. That said, no idea is a bad idea and every idea needs considering – including the trivial ones. Brainstorming sessions need to be well managed.

- **Mind mapping** – a useful technique for summarising and organising a lot of information into a visual diagram, helping you to identify the areas to consider, spot what is missing and create connections.

One way to evaluate and improve the user stories is to use a quality check like the INVEST acronym:

Independent	Should not have any dependencies with other user stories.
Negotiable	Should provide just enough information to enable later negotiation around elaboration, clarification and prioritisation.
Valuable	Should provide details of the potential business value.
Estimatable	Should have enough information to enable the team to estimate the amount of work needed to deliver it.
Small	Should be small enough to go from being a user story to potentially shippable code within the one sprint/iteration.
Testable	Should have specific measures detailed to enable it to be tested and evaluated against the acceptance criteria.

If the user story does not comply with the INVEST acronym, this may mean that the stakeholders don't know how to test it, or may not fully understand what the user story is trying to achieve, or it may not be clear enough or it could be vague. The testers should get involved with helping to define the acceptance criteria as they will need to ensure that each criterion is testable and can be confirmed with a definite 'yes' or 'no' as to whether it has been met following test execution.

Acceptance criteria

Acceptance criteria are the measures or conditions defined to evaluate and determine whether the need, requirement or user story are acceptable to the stakeholders. They usually spring from the results of the conversation held as part of the 3Cs and, from a testing point of view, these criteria help to define the test conditions to execute.

Acceptance criteria determine the scope of the user story, gain agreement between the stakeholders as to what the user story includes, cover both positive and negative scenarios, enable the user story to be included in further planning and estimating activities, and can be used as a basis for acceptance test-driven development (see the next section).

There are several different formats available to check against the acceptance criteria. Two of the most common are:

Scenario-oriented: Usually in the given–when–then format, these can elaborate the acceptance criteria and include all information needed to use this as a test.

Rule-oriented: Usually a bullet list or table of items to check against the acceptance criteria. May not have as much information as a scenario.

EXAMPLE 14: WITHDRAW CASH FROM AN AUTOMATED TELLER MACHINE (ATM)

We will use an example of withdrawing £50 from an ATM. We have a user story defined as follows:

> As a current account holder,
> I want to withdraw £50 from an ATM,
> So that I don't need to go into the bank to get to my money out.

Looking at this user story, we would first need to identify the positive acceptance criteria where we are able to withdraw £50 from the ATM. Rule-oriented criteria may be as follows:

- The machine will produce £50 in notes and return the card when taking £50 from an account holding at least £50 and the machine has the correct denominations of notes.

The above have so far been functional acceptance criteria; we may need to also consider non-functional acceptance criteria:

- The end-to-end response time of the ATM needs to be less than three seconds (performance efficiency).

- The ATM must output the notes that make £50 within five seconds of pressing the Withdraw button (performance efficiency).

- The user must be able to withdraw cash without needing to ask any of the staff inside the bank how to do this (usability).

We can then look at the negative acceptance criteria, which could be the opposite of the positives:

- If there is less than £50 in the current account, display a message stating that you cannot withdraw this much and state how much is in the account to possibly withdraw.

- If the machine does not hold £50 in notes, display a message stating that the machine does not have enough money and state the highest amount in the machine that is withdrawable.

- If the machine does not have the right denominations of notes to make £50, display a message stating that you can only withdraw cash in multiples of the lowest denomination note available.

Acceptance test-driven development

Acceptance test-driven development (ATDD; see **Figure 4.75**) is an example of applying a shift-left approach, as discussed in **Chapter 2**. In traditional sequential projects the code is written first and then the tests are designed and executed, which means you are checking your expected results against the already (code) generated actual results. In ATDD, the tests, which include the expected results, are designed **before** the code is written and therefore **before** the actual results are generated from code execution. Ideally, a team of people with different perspectives (e.g. customers, developers, testers) get together to write the test cases to show that we have considered all angles of the user story.

Following the 3Cs, a user story is created that may or may not have some acceptance criteria. The team analyse and discuss the user story with a view to understanding what it is trying to achieve. At this point they may find and resolve defects or vague ambiguities or gaps. They also help to generate, define or refine the acceptance criteria.

After defining and agreeing the user story and acceptance criteria, the next step is to generate test cases to cover the acceptance criteria. These test cases should use language everyone understands (including the stakeholders, developers and testers) and ideally be expressed in a natural language covering preconditions, inputs, expected outputs and postconditions (in other word, the parts of a test case – see the 'Test design' section of this chapter for more details). The test cases need to cover all the acceptance criteria, there should be no gaps, no duplicates and they shouldn't go beyond the scope of the user story.

Figure 4.75 Acceptance test-driven development

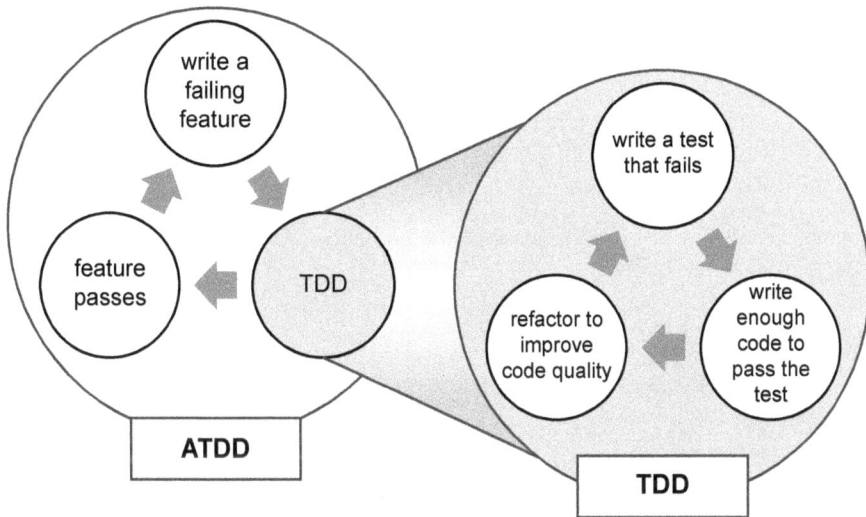

The test cases could be more high-level examples of how the acceptance criteria would work; for example:

- Preconditions:
    - account holder has £100 in their account;
    - machine holds the right notes to dispense £50.
- Input: account holder requests £50.
- Expected results: £50 dispensed and card returned by the machine.

Test cases could also be designed by applying test techniques (e.g. three-value BVA on the £50 boundary would create test cases for £45, £50, and £55 as the lowest denomination note from an ATM is £5[8]).

Ideally, the first test cases would cover the positive functional scenarios where everything works and behaves correctly as expected, with the next set of test cases addressing negative functional testing. The final set of test cases would check other quality characteristics, usually non-functional (e.g. usability, reliability, performance efficiency, etc.).

Once written, the test cases would be executed manually or through automation. If automated, then either a developer or technical test analyst converts the steps into code that is included in a test automation framework. In this way, the scenarios created from the acceptance criteria become executable specifications whereby they document how the user story works, and are also used for execution and reporting.

---

8 At the time of writing.

REVISITING EXAMPLE 14: WITHDRAW CASH FROM AN ATM

We reuse the ATM example from the previous section with the following user story defined:

> As a current account holder
> I want to withdraw £50 from an ATM
> So that I don't need to go into the bank to get to my money out.

The first acceptance criteria to identify is the positive outcome. Scenario-oriented criteria may be as follows:

> Given I have at least £50 in my account
> And the machine has the correct note denominations
> When I withdraw £50 from the machine
> Then the machine will give me £50
> And return my card.

We will need to also consider non-functional acceptance criteria:

> Given I have requested to withdraw £50
> And there is enough money in my account
> And the machine has the correct note denominations
> When the machine is processing my request
> Then the machine will take no longer than 5 seconds to produce the money
> And return my card

> Given I am interacting with the machine
> When I press any of the buttons on the machine
> Then the machine will take no longer than 3 seconds to produce a response

The negative acceptance criteria would need to include opposites of the positive acceptance criteria:

> Given I have requested to withdraw £50
> And there is less than £50 in my account
> When I attempt to withdraw £50 from the machine
> Then the machine will display a message stating I do not have enough money
> And shows how much is in the account to possibly withdraw.

> Given I have requested to withdraw £50
> And there is enough money in my account
> And the machine does NOT have the correct note denominations
> When I attempt to withdraw £50 from the machine
> Then the machine will display a message stating that you can only withdraw cash in multiples of the lowest denomination note available.

CHECK OF UNDERSTANDING

1. What are the 3Cs?

2. What are the two common formats for acceptance criteria?

3. What is the next step in ATDD after discussing the user story and generating the acceptance criteria?

SUMMARY

In this chapter we initially looked at the terminology and test work products created as part of the test development process, identifying how test conditions are created during test analysis, test cases during test design, and test procedures during test implementation in readiness for test execution.

We also looked at how test techniques are grouped into three categories: black box (or specification-based), white box (or structure-based) and experience-based. The black box test techniques covered were: equivalence partitioning, boundary value analysis, state transition testing and decision table testing. White box test techniques covered were statement testing and branch testing, with an explanation showing how the white box test techniques have a hierarchy and how one technique 'subsumes' the technique below it. Experience-based test techniques we discussed were error guessing, exploratory testing and checklist-based testing.

We also looked at collaboration-based test approaches (rather than test techniques) and considered the 3Cs (card, conversation and confirmation) within collaborative user story writing as well as looking into defining acceptance criteria. Finally, we discussed acceptance test-driven development and how acceptance criteria are included in this approach.

Example examination questions with answers

E1. K2 question
Which of the following correctly characterises white box test techniques?

A. Test cases may be used to detect differences between requirements and implementation.

B. Test cases may be used to determine deviations from requirements.

C. Test cases may be based on software architecture and used to exercise interfaces.

D. Test cases may be based on user stories and used to exercise use cases.

E2. K2 question
Which of the following identifies a key difference between black box and white box test techniques?

    A. Coverage measures are applied to white box test cases but not to black box test cases.

    B. Black box test cases may be based on requirements or on the tester's experience; white box test cases are based on structure.

    C. Coverage measures are applied to black box test cases but not to white box test cases.

    D. Black box tests can determine deviations from requirements; white box tests can determine deviations from design.

E3. K3 question
A washing machine has three temperature bands for different kinds of fabrics: fragile fabrics are washed at temperatures between 15 and 30 degrees Celsius; normal fabrics are washed at temperatures between 31 and 60 degrees Celsius; heavily soiled and tough fabrics are washed at temperatures between 61 and 100 degrees Celsius. Which of the following contains only values that are in different equivalence partitions?

    A. 15, 30, 60.

    B. 20, 35, 60.

    C. 25, 45, 75.

    D. 12, 35, 55.

E4. K3 question
The following state transition diagram represents a series of valid transitions between five states.

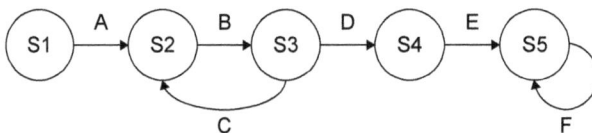

Which of the following test cases is the minimum number of state transitions needed to provide 100 per cent all states coverage?

    A. A, B, D, E.

    B. A, B, D, E, F, F.

    C. A, B, C, B, D, E, D.

    D. A, B, C, B, D, E, F.

E5. K3 question

The following state transition diagram represents a series of valid transitions between five states.

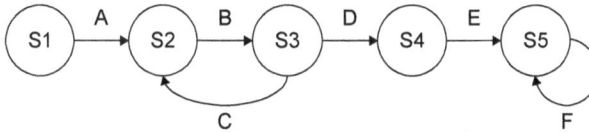

Which of the following test cases is the minimum number of state transitions needed to provide 100 per cent valid transitions coverage?

    A. A, B, D, E.
    B. A, B, D, E, F, F.
    C. A, B, C, B, D, E, D.
    D. A, B, C, B, D, E, F.

E6. K3 question

A pricing system for a mountain railway has three price tariffs:

- morning tariff – for journeys between 6 a.m. and 12 noon inclusive;
- afternoon tariff – for journeys up to 4 p.m. inclusive;
- evening tariff – for journeys up to 8 p.m. inclusive.

The service does not operate outside the times shown above (i.e. there is no night service).

Customers must enter their preferred journey time in a data entry field, in the format 'hh:mm' (24-hour clock).

Which of the following test suites provide the best equivalence partition coverage for the system, considering valid and invalid equivalence partitions?

    A. 06:00, 11:59, 15:00, 19:00.
    B. 05:00, 07:00, 18:00, 23:00.
    C. 00:30, 09:00, 14:30, 18:00.
    D. 04:30, 08:30, 12:30, 16:00.

E7. K3 question

	Rule 1	Rule 2	Rule 3	Rule 4	Rule 5	Rule 6
CONDITIONS						
Diamond member	T	F	F	F	F	F
Gold member	-	T	T	F	F	F
Silver member	-	F	F	T	T	F
Long haul	-	T	F	T	F	-
ACTIONS						
Express check-in	YES	YES	YES	NO	NO	NO
In-flight meal charge	FREE	FREE	FREE	£10	£20	£25
Extra baggage charge per kilo	FREE	£20	£10	£30	£15	Extra baggage not allowed

The decision table above reflects an airline's pricing scheme. What are the expected results for the following two passengers:

- Paul is a Gold Card holder flying short haul and has 2 kg extra baggage.

- Cheryl is a Silver Card holder flying long haul and has 1 kg extra baggage.

    A.   Paul pays £10 for extra baggage but gets express check-in and a free in-flight meal; Cheryl pays an extra £40 for both a meal and extra baggage, but isn't entitled to express check-in.

    B.   Paul pays £20 for extra baggage but gets express check-in and a free in-flight meal; Cheryl pays an extra £40 for both a meal and extra baggage, but isn't entitled to express check-in.

    C.   Paul pays nothing for extra baggage and gets express check-in and a free in-flight meal; Cheryl pays an extra £25 for a meal but isn't entitled to express check-in and can't take on extra baggage.

    D.   Paul pays £20 for extra baggage but gets express check-in and a free in-flight meal; Cheryl pays an extra £35 for both a meal and baggage but isn't entitled to express check-in.

Answers to the self-assessment questions in the chapter

SA1. The correct answer is B.

SA2. The correct answer is A.

SA3. The correct answer is B.

Exercise 4.1
The partitions are: £0.00–£1,000.00, £1,000.01–£2,000.00 and ≥£2,000.01.

Exercise 4.2
The valid partitions are: £0.00–£20.00, £20.01–£40.00 and ≥£40.01. Non-valid partitions would include negative values and non-numeric characters.

Exercise 4.3
The partitions are: question scores 0–20; total 0–100; question differences: 0–3 and >3; total differences 0–10 and >10.

Boundary values are: –1, 0, 1 and 19, 20, 21 for the question scores; –1, 0, 1 (again) and 99, 100, 101 for the question paper totals; –1, 0, 1 (again) and 2, 3, 4 for differences between question scores for different markers; and –1, 0, 1 (again) and 9, 10, 11 for total differences between different markers.

In this case, although the values –1, 0, 1 occur several times, they may be applied to different parts of the program (e.g. the question score checks will probably be in a different part of the program from the total score checks) so we may need to repeat these values in the boundary tests.

Exercise 4.4
Joyce will be eligible for a cash payment but not for a share allocation.

Answers to example examination questions

E1. The correct answer is C.

White box techniques are about the structure of software solutions, so they typically exercise interfaces and check that the software architecture has been correctly implemented. They do not reference requirements as such, and therefore cannot determine deviations from requirements or whether requirements have been correctly implemented. User stories are a form of requirements, so white box testing is not appropriate to user stories or use cases.

E2. The correct answer is D because black box testing is based on requirements and white box testing is based on design.

      A. is incorrect because coverage measures can be applied to both black box and white box tests.

      B. is partially correct, in that white box test cases are based on structure and black box test cases are based on requirements but not the tester's experience. Experience-based testing is based on the tester's experience.

      C. is incorrect for the same reason as option A is incorrect.

E3. The correct answer is C.

      A. includes two values from the lower partition;

      B. contains two values from the second partition;

      D. contains one value that is invalid (out of range).

E4. The correct answer is A.

'All states' coverage just covers all states on the diagrams. It does **not** need to cover all the transitions. The test case that covers the states S1 to S5 in the minimum number of transitions is option A. The other answers will also provide 100 per cent all states coverage but are not the minimum number needed.

E5. The correct answer is D. 'Valid transitions' coverage covers all valid transitions on the diagram (and therefore also all the states).

    A.   provides 100 per cent all states coverage but does not cover all transitions.

    B.   covers all transitions apart from transition C.

    C.   covers all transitions apart from transition F.

E6. The correct answer is C.

The partitions in hh:mm 24-hour format are as follows:

Morning   06:00–12:00
Afternoon   12:01–16:00
Evening   16:01–20:00
'Nighttime'   20:01–05:59

Nighttime is in quotation marks as the question does not specifically state this time range but needs to be included when considering 100 per cent equivalence partition coverage.

    A.   covers three partitions: Morning twice (06:00 and 11:59), Afternoon (15:00) and Evening (19:00) but does not cover Nighttime.

    B.   covers three partitions: Morning (07:00), Evening (18:00) and Nighttime twice (05:00 and 23:00) but does not cover Afternoon.

    D.   covers three partitions: Nighttime (04:30), Morning (08:30) and Afternoon twice (12:30 and 16:00) but does not cover Evening.

E7. The answer is B.

Paul is a Gold Card holder ('Diamond member' is F, 'Gold member' is T, 'Silver member' is F) and is flying short haul ('Long haul' is F), which makes him Rule 3. The actions for Rule 3 show 'Express check-in' as YES, 'In-flight meal charge' as FREE and he has 2 kg extra baggage, which means he has two times the 'Extra baggage charge per kilo', which is 2 × £10 = £20.

Cheryl is a Silver Card holder ('Diamond member' is F, 'Gold member' is F, 'Silver member' is T) and is flying long haul ('Long haul' is T), which makes her Rule 4. The actions for Rule 4 show 'Express check-in' as NO (she does not get Express check-in), the 'In-flight meal charge' is £10 and she has 1 kg extra baggage and 'Extra baggage charge' is £30. Adding the meal and baggage charges together makes £40.

5 TEST MANAGEMENT

Geoff Thompson

INTRODUCTION

This chapter provides an overview of how test activities are managed in sequential and iterative software development life cycle (SDLC) models.

We will start by looking at the activity of test planning, including the definition of entry and exit criteria, techniques for estimating the testing effort, how to prioritise tests and the concepts behind the test pyramid and testing quadrants. We will discuss how testing and risk fit together, how we monitor and control the test activities and how we communicate the status of testing to the stakeholders. We further discuss how configuration management supports testing and how to prepare well documented defect reports.

Learning objectives

The learning objectives for this chapter are listed below. You can confirm that you have achieved these by using the self-assessment questions immediately following the learning objectives, the 'Check of understanding' boxes distributed throughout the text and the example examination questions provided at the end of the chapter. The chapter summary will remind you of the key ideas.

Each learning objective is allocated a K number to represent the level of understanding required; see the **Introduction** (pp. 2–3) for an explanation of K numbers.

Test planning

- FL-5.1.1 (K2) Exemplify the purpose and content of a test plan.
- FL-5.1.2 (K1) Recognise how a tester adds value to iteration and release planning.
- FL-5.1.3 (K2) Compare and contrast entry criteria and exit criteria.
- FL-5.1.4 (K3) Use estimation techniques to calculate the required test effort.
- FL-5.1.5 (K3) Apply test case prioritisation.
- FL-5.1.6 (K1) Recall the concepts of the test pyramid.
- FL-5.1.7 (K2) Summarise the testing quadrants and their relationships with test levels and test types.

Risk management

- FL-5.2.1 (K1) Identify risk level by using risk likelihood and risk impact.
- FL-5.2.2 (K2) Distinguish between project risks and product risks.

- FL-5.2.3 (K2) Explain how product risk analysis may influence thoroughness and scope of testing.
- FL-5.2.4 (K2) Explain what measures can be taken in response to analysed product risks.

Test monitoring, test control and test completion

- FL-5.3.1 (K1) Recall metrics used for testing.
- FL-5.3.2 (K2) Summarise the purposes, contents and audiences for test reports.
- FL-5.3.3 (K2) Exemplify how to communicate the status of testing.

Configuration management

- FL-5.4.1 (K2) Summarise how configuration management supports testing.

Defect management

- FL-5.5.1 (K3) Prepare a defect report.

Self-assessment questions

The following questions have been designed to enable you to check your current level of understanding for the topics in this chapter. The answers are at the end of the chapter.

Question SA1 (K1)
Which of the following is a valid exit criterion from the test execution phase?

    A.   All tests have been defined.
    B.   All defects reported have been corrected.
    C.   All planned tests have been executed.
    D.   All testing tasks have been assigned.

Question SA2 (K2)
How might configuration management support testing?

    A.   By recording the versions of all items that make up a test environment.
    B.   By tracking the status of all unresolved high-severity defect reports.
    C.   By calculating the coverage achieved by a suite of executed unit tests.
    D.   By determining which test cases are best suited for automation.

Question SA3 (K1)
What can a risk-based approach to testing provide?

    A.   The types of test techniques to be employed.

    B.   The total tests needed to provide 100 per cent coverage.

    C.   An estimation of the total cost of testing.

    D.   Only that test execution is effective at reducing risk.

TEST PLANNING

Test plan

Test planning is a key activity in any test project. The results of the test planning activity are documented in a test plan. ISO/IEC/IEEE 29119-3 *Software and Systems Engineering – Software Testing – Part 3: Test Documentation* defines the content of a test plan.

The test plan:

1. documents the schedule of activities needed to achieve the test objectives;

2. documents what is required (the means) to achieve the test objectives – for example, the number of test resources needed to complete the scheduled activities, the test environment requirements, what is in and out of scope, the approach to testing, the test strategy (if required), entry and exit criteria and the risks;

3. provides a method of communicating the test activities to be performed to all team members and stakeholders;

4. ensures that the test activity is agreed as the best way to achieve the criteria defined (test activity is any activity involved in delivering the testing required; this could include team meetings, test design, test execution and more);

5. demonstrates how testing will comply with the test policy and corporate/project test strategy or confirms the reasons why it doesn't and identifies any additions/ deletions.

Test planning should be initiated as early in the project life cycle as possible to guide the testers' thinking when they start test development and force the testers and the project to confront the future challenges in the project, in areas such as risk, schedules, people, tools, costs and effort. Test managers find the preparation of a test plan a useful tool to ensure they consider all eventualities and enables them time to think through what is required to achieve the test objectives.

In large projects a master test plan or a project test plan may be produced. This document defines at a high level the test activities being planned. It is normally produced during the early phases of the project (e.g. initiation) and updated as required via change control as the project develops. It will provide sufficient information to enable a test project to be established (bearing in mind that at this point in a project little more than requirements may be available from which to plan).

It includes the details of the test-level activities and defines the test-level plans to be developed, such as the system test plan. These documents will contain the detailed activities and estimates for the relevant test level (e.g. system test) or test types (e.g. security and performance testing).

Figure 5.1 shows where test-level test plans fit into the V model. It shows that a test plan exists for each test level and that they will usually refer to the master test plan.

Figure 5.1 Test plans in the V model

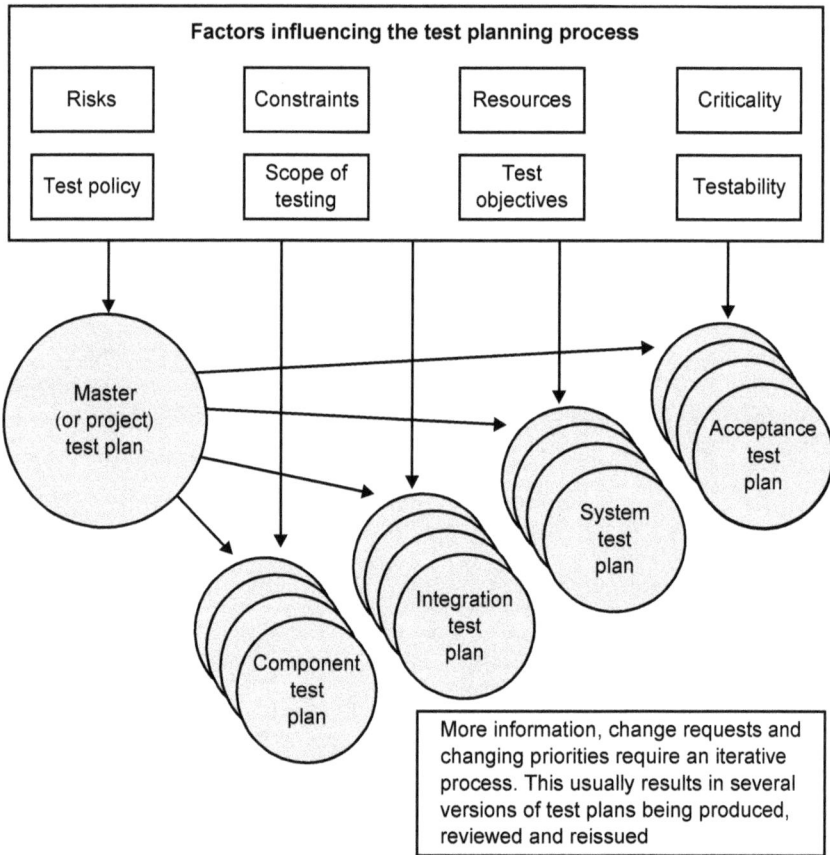

The contents sections of a test plan for either the master test plan or test-level plans are normally identical or very similar. The typical content of a test plan includes the following:

- Context of testing – this includes the scope, what the test objectives are (e.g. achieve 90 per cent requirements coverage) and what the test basis is (what is to be used to provide the details against which tests will be built – e.g. a requirements document, or a requirements backlog).

- Assumptions and constraints impacting tests and the test project.
- Stakeholders – this includes a list of all roles involved (test and defect management roles) including any dependencies for testing (e.g. the test will have a dependency on the use of a test environment, so who is the contact that ensures the team get the environment service they need?). For each role, a set of responsibilities defining what is expected of the person in the role. Finally, specific training for team members to enable them to complete their responsibilities, as well as any hiring needs where resources are not available from the existing resource pool.
- Approach to communication – what will be communicated and when and to whom, and any templates to be used.
- A risk register documenting all project and product risks for testing alongside any mitigating actions required.
- The test approach, which will explain:
  - test levels;
  - test types;
  - test techniques to be used;
  - test deliverables;
  - entry and exit criteria from both test design and test execution;
  - the independence of testing;
  - metrics to be collected;
  - test data requirements;
  - test environment requirements;
  - deviations from the test policy or test strategy.
- Budget requirements and the test schedule.

During test planning various activities for an entire system or a part of a system have to be undertaken by those working on the plan. These include the following:

- Working with the project manager and subject-matter experts to determine the scope and the risks that need to be tested. Also, identifying and agreeing the objectives of the testing, be they time-, quality- or cost-focused, or a mixture of all three. The objectives will enable the test project to know when it has finished – has time or money run out, or has the right level of quality been met?
- Understanding what delivery model is to be used (waterfall, iterative, Agile, etc.) and defining the overall approach of testing (sometimes called the test strategy) based on this, ensuring that the test levels and entry and exit criteria are defined.
- Liaising with the project manager and making sure that the testing activities have been included within the software life cycle activities such as:
  - design – the development of the software design;
  - development – the building of the code;

- implementation – the activities surrounding implementation into a live environment.

- Working with the project to decide what needs to be tested, what roles are involved and who will perform the test activities, planning when and how the test activities should be done, deciding how the test results will be evaluated, and defining when to stop testing (exit criteria).

- Building a plan that identifies when and who will undertake the test analysis and design activities. In addition to the analysis and design activities, test planning should also document the schedule for test implementation, execution and evaluation. The plan can either be sequential (e.g. particular dates are defined) or iterative, where the context of each iteration will need to be considered.

- Deciding what the documentation for the test project will be (e.g. which plans, how the test cases will be documented and so on).

- Defining the management information, including the metrics required, and putting in place the processes to monitor and control test preparation and execution, defect resolution and risk issues.

- Ensuring that the test documentation generates repeatable test assets (e.g. test cases).

Testers' contributions to iteration and release planning

During iterative SDLCs there are typically two types of planning that could take place: release planning and iteration planning.

Release planning looks ahead to when the product is released, it defines and redefines the product backlog. If the user stories are large and cumbersome, during release planning they may be refined into a set of smaller user stories.

Once completed, release planning becomes the basis upon which the test approach and test plan is developed across all iterations.

Testers play a critical part in release planning. They assist in the writing of testable user stories and associated acceptance criteria (see **Chapter 4**), assist in the project and product risk analysis and help to refine the test estimate associated with running test stories. They can also be involved in the development of the test approach and the release planning itself.

Iteration planning looks ahead at the planning of a single iteration and focuses on that iteration's backlog definition.

Testers will become involved in detailed risk analysis of user stories, they will check the testability of user stories and look to break user stories down into tasks that will include testing tasks. Using what they discover, they estimate the testing task effort and ensure all functional and non-functional aspects of the test object have been identified and, if necessary, refined/corrected/updated.

Entry criteria and exit criteria

Entry criteria (also known as Definition of Ready in Agile) are used to determine the preconditions to be met before a given test activity can start. This could include the start of test planning and when test design and test execution for each level of testing is ready to start.

Examples of some typical entry criteria to test execution may include:

- people required to run the tests;
- sufficient budget/time is available;
- availability of testware such as test cases, test procedures and the test execution schedule;
- test environment is available and ready for use (it functions);
- test tools installed in the environment are ready for use;
- testable code is available;
- all test data is available and correct;
- the previous test activity has completed and met its exit criteria (e.g. all smoke tests have passed).

Exit criteria (also known as the Definition of Done in Agile) are used to determine when a given test activity has been completed or when it should stop. Exit criteria can be defined for all of the test activities, such as planning, specification and execution as a whole, or to a specific test level for test specification, as well as execution.

Some typical exit criteria might be:

- Measures of thoroughness:
    - A certain level of coverage has been achieved.
    - The number of unresolved defects is within an agreed limit.
    - Defect density.
    - Number of failed tests.
- Completion criteria:
    - All tests planned have been executed.
    - All high-risk areas have been fully tested, with only minor residual risks left outstanding.
    - Cost – when the budget has been spent.
    - All defects found have been reported.
    - All regression tests have been automated.
    - The evaluated level of quality criteria, such as reliability and performance, is sufficient.

■ The schedule has been achieved – for example, the release date has been reached and the product has to go live. The release date can be set externally, such as the millennium testing (it had to be completed before midnight on 31 December 1999) and is often the case with government legislation, or it can be set by the project and once committed to must be met.

Entry and exit criteria should have been agreed as early as possible in the life cycle; however, they can be, and often are, subject to controlled change as the detail of the project becomes better understood and therefore the ability to meet the criteria is better understood by those responsible for delivery.

CHECK OF UNDERSTANDING

1. What is the international standard for software testing called?
2. Give three examples of exit criteria.
3. What activities are contained within test planning?
4. Detail four areas that a tester would support in iteration planning.

Test estimation

Test estimation involves predicting the number and length of tasks involved in meeting the objectives of the test within the project. Estimates will, by nature, be based on some assumptions that should be declared along with the estimate. The larger the task the more chance of error in the estimate – therefore, it is often better to split the estimate up into smaller component parts as small project estimates are known to be more accurate.

Within the syllabus there are four approaches to test estimation referenced: estimate based on ratios, extrapolation, Wideband Delphi and three-point estimation.

Estimate based on ratios
This is a metrics-based approach and relies on data collected from previous or similar projects that can be used to define a standard ratio for similar projects. This kind of data might include:

- the number of test conditions;
- the number of test cases written;
- the number of test cases executed;
- the time taken to develop test cases;
- the time taken to run test cases;
- the number of defects found;
- the number of environment outages and how long on average each one lasted.

With this approach and the right data, it is possible to estimate quite accurately what the cost and time required for a similar project would be.

It is important that the actual costs and time for testing are accurately recorded. These can then be used to revalidate and possibly update the metrics for use on the next similar project.

For example, if a project of a similar type and size created 300 test cases, you could estimate that the current project would create roughly the same amount. Another example could be that a project had a development to test ratio of 3:1, and that that project delivered near-perfect software. Therefore, the next similar project could use this to estimate the rough testing cost. Assuming the new project development was 600 days of effort, testing can be estimated at 200 working days of effort.

Extrapolation
This is another metrics-based approach to estimating, which involves making measurements as early as possible in the current project to enable data to be derived upon which the estimate will be based. When sufficient data is collected, the effort for the remaining work can be derived normally using a mathematical method. An example of this is using the data collected in the first three iterations of a project to estimate the average that can be used as the estimate for the next iteration (e.g. iteration 1 spent 100 hours on test, iteration 2 spent 120 hours on test and iteration 3 spent 98 hours on test, therefore the average of the three iterations is 106 hours, which could be the estimate for iteration 4).

Wideband Delphi
This expert-based approach to metrics uses the experience of owners of the relevant tasks or experts in a particular domain to derive an estimate (often called the 'Expert' method). In this context, 'experts' could be:

- business experts;
- test process consultants;
- developers;
- technical architects;
- analysts and designers;
- anyone with knowledge of the application to be tested or the tasks involved in the process.

The approach to developing the estimate would involve distributing a requirement specification (or other test basis) to the experts and getting them to estimate their task in isolation. Assuming there is consensus between the individual estimates, you can build in any required contingency to arrive at the estimate.

However, where there are wide differences between estimates, the experts, again in isolation, must revisit their estimates and only when there is consensus can the process complete.

Planning Poker is a variant of Wideband Delphi used in Agile projects. This approach uses a pack of cards with numbers, often based upon the Fibonacci sequence, that represent effort sizes. The team will discuss the scope and effort of each user story as a team, then compare everyone's anonymous estimates provided as a number on a card. If the team doesn't come to a consensus, they will have a discussion to better understand the work, then estimate again until the team reaches an agreement.

Three-point estimation

This is another expert-based technique. The expert is asked to produce three estimates: (a) the most optimistic, (m) the most likely and (b) the most pessimistic. Then, using the following calculation method, an estimate can be calculated as:

$$estimate = (a + (4 \times m) + b)/6.$$

To see this in action, the three estimates that the expert provided were:

- a (the most optimistic) = 120 working days
- m (the most likely) = 150 working days
- b (the most pessimistic) = 200 working days

The calculation of the estimate would then be:

(120 + (4 × 150) + 200)/6 = 153 working days

To calculate the margin of error the expert will use the calculation (b – a) / 6:

(200 – 120)/6 = 13 working days

Therefore, the estimate would be:

153 man days ± 13 man days, so 140 to 166 working days

In support of the estimation process, many things affect the level of effort required to fulfil the test-related elements of a project and would need to be considered when estimating the test effort. These can be split into four main categories:

1. Product characteristics:
   - The risks associated with the product (defined during risk-based testing).
   - The quality of the test basis – for example, requirements, user stories and so on.
   - The complexity of the product domain.
   - The number of quality characteristics, such as reliability.
   - The number of non-functional requirements.
   - The security requirements (perhaps meeting ISO 27001, the security standard).

- How much documentation is required (e.g. some legislation-driven changes demand a certain level of documentation that may be more than an organisation would normally produce).

- Requirements for legal or regulatory compliance.

2. Development process characteristics:

- The stability and maturity of the organisation; for example, a very process-mature organisation will take a lot less time to achieve what an immature (seat of their pants) organisation would take, as they are likely to make fewer mistakes.

- The development model in use, such as Agile or sequential.

- The agreed test approach.

- The tools in use, automation, test management, etc.

- The test process defined in the test strategy and approach.

- Timescales.

3. People characteristics:

- The skills of those involved in the testing and development activity (the lower the skill level in development, the more defects could be introduced, and the lower the skill level in testing, the more detailed the test documentation needs to be).

- Team cohesion and leadership.

4. Test results:

- The number and severity of defects expected to be found.

- The amount of rework needed.

CHECK OF UNDERSTANDING

1. Compare and contrast the two approaches to developing estimates: metrics and expert.

2. Provide three examples of what a metrics approach to estimates would use as a base.

3. Name three experts that could be involved in Wideband Delphi estimation.

Test case prioritisation

To ensure that test cases and test procedures are run in order of importance they need to be prioritised. Once identified, the most important is executed first. If no formal approach to prioritisation is used, test cases and test procedures will be run in a random order, which may or may not be the right order for the project/iteration. Strategies for test case prioritisation include the following:

- Risk-based prioritisation – the order of test execution is defined based on the output of the risk analysis of each test, where the test cases or test procedures are assessed, and the likelihood and impact of test failure will be understood. The tests with the highest risk level (probability and impact) would be run first to ensure that the defects are found as early as possible in the test execution cycle.

- Coverage-based prioritisation – the order of test execution is judged based on the coverage achieved by the test cases or test procedures. The test cases or test procedures that provide the highest coverage will be run first, thus achieving higher test coverage as quickly as possible.

- A variant of coverage-based prioritisation is called **additional coverage prioritisation** and involves the test case or test procedure with the highest coverage being run first, and then all subsequent test cases or test procedures being prioritised by the amount of additional coverage they provide.

- Requirements-based prioritisation – when the requirements have been prioritised by the stakeholders (e.g. product owner), the test cases that are traceable to the high-priority requirements are executed first.

In many situations, even when prioritised, test cases and test procedures may not be run in the priority order due to other dependencies within the test case or procedure. For example, you can't cancel a product before it has been created, even if the test execution for cancellation is the most important – the tester will first have to set a product up on the system (if it doesn't exist on the system already), which may be a low-priority test but has to be run first.

An additional impact on prioritisation is the availability of the test environment. It may not have the required software installed so that a test can be run, or resources to carry out execution are not available, or tools needed to find the required test data may not yet be available.

In general, prioritisation can become quite complex when you also consider the dependencies that exist within a test cycle, but it is key that everyone understands and agrees the order in which test cases or test procedures are executed.

Test pyramid

Developed by Cohn in 2009, the test pyramid (also known as the test automation pyramid) is a model that shows that when automating tests, different test types may have different granularity (**Figure 5.2**).

Essentially, the testing pyramid lays out the types of tests that should be included in an automated test suite. It also outlines the sequence and frequency of these tests.

The traditional test pyramid references three groups of tests:

- Unit tests form the base of the test automation pyramid. They test an individual component or function to validate that it works as expected in isolated conditions. It is essential to run several types of scenarios in unit tests – happy path (a situation in which all tests run as they are supposed to, with no faults), error handling, etc.

Figure 5.2 The test pyramid

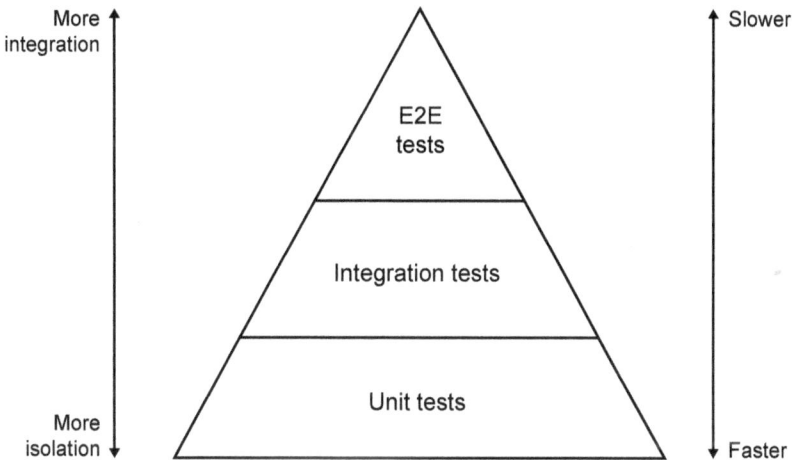

- Since this is the most significant subset, the unit test suite must be written to run as quickly as possible.

- Remember that the number of unit tests will increase as more features are added.

- This test suite must be run every time a new feature is added.

- Consequently, developers receive immediate feedback on whether individual features are working.

A fast-running unit test suite encourages developers to run it as often as possible. An excellent way to build a robust unit test suite is to practice test-driven development (TDD). Since TDD requires a test to be written before any code, the code is more straightforward, transparent and bug-free.

- Integration tests need to be run to test how each unit of code interacts with other code (that will form the entire software release). Essentially, these are tests that validate the interaction of a piece of code with external components. These components can range from databases to external services (application program interfaces (APIs)).

  - Integration tests are the middle layer of the test automation pyramid. This means that they will not be run as frequently as unit tests.

  - Fundamentally, they test how a feature communicates with external dependencies.

  - Whether a call to a database or web service, the software must communicate effectively and retrieve the correct information to function as expected.

- The top level of the testing pyramid is the end-to-end tests. These ensure that the entire application is functioning as required. End-to-end tests do precisely what the name suggests: they test that the application works flawlessly from start to finish.

- End-to-end tests are at the top of the testing pyramid as they take the longest to run.

- When running these tests, it is essential to imagine the user's perspective.

- How would an actual user interact with the app? How can tests be written to replicate that interaction?

They can also be fragile and therefore break very easily since they have to test various user scenarios. Like integration tests, these tests may also require the app to communicate with external dependencies, thus adding to possible bottlenecks in completion.

Test quadrants

Developed by Brian Marick, testing quadrants are a visual tool for understanding different test types used during Agile development. The quadrants differentiate between business and technology-facing tests, and those that support programming or 'critique' the product. Testing types are sorted into four categories on a grid. The grid aids test management in visualising the tests to ensure that all test types and test levels have been included in the SDLC plan.

The four quadrants (**Figure 5.3**) are:

Figure 5.3 The testing quadrant

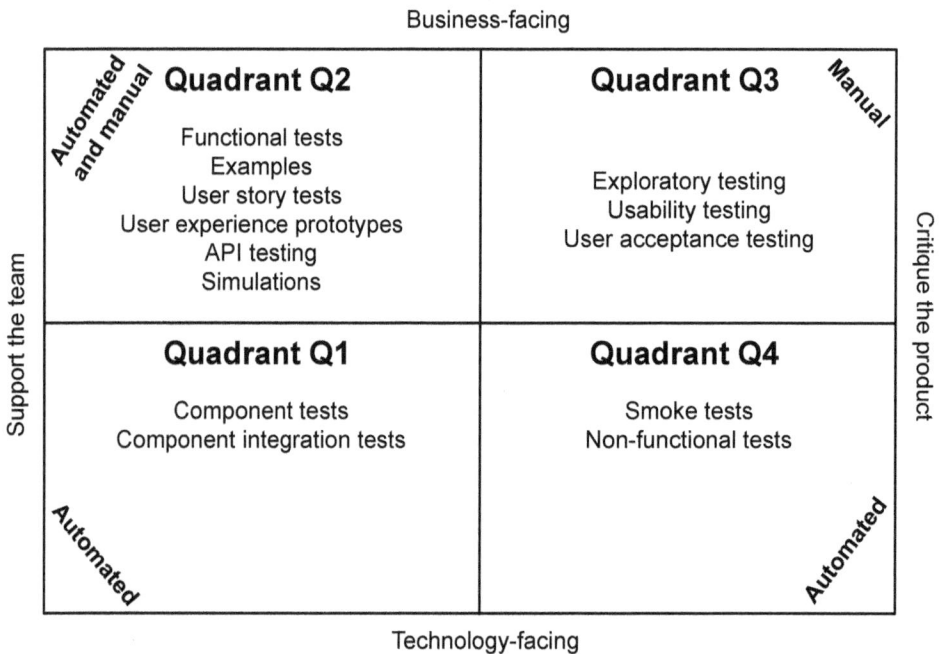

Business-facing

Quadrant Q2 *(Automated and manual)*	Quadrant Q3 *(Manual)*
Functional tests Examples User story tests User experience prototypes API testing Simulations	Exploratory testing Usability testing User acceptance testing
Quadrant Q1 *(Automated)*	Quadrant Q4 *(Automated)*
Component tests Component integration tests	Smoke tests Non-functional tests

Support the team   Critique the product

Technology-facing

- Quadrant Q1 (technology-facing, support the team). This quadrant contains component and component integration tests. These tests should be automated and included in the process.

- Quadrant Q2 (business-facing, support the team). This quadrant contains functional tests, examples, user story tests, user experience prototypes, API testing and simulations. These tests check the acceptance criteria and can be manual or automated.

- Quadrant Q3 (business-facing, critique the product). This quadrant contains exploratory testing, usability testing and user acceptance testing. These tests are user-oriented and often manual.

- Quadrant Q4 (technology-facing, critique the product). This quadrant contains smoke tests and non-functional tests (except usability tests). These tests are often automated.

The testing quadrant provides a holistic overview of software testing. It can inform your decision making, for example in answering the question: Have we covered all the bases?

RISK MANAGEMENT

It is not possible to talk about test management without looking at risk and how it affects a generic test process as defined in **Chapter 1**. If there were no risk of adverse future events (defects) in software or hardware development, then there would be no need for testing. In other words, if the risk of defects did not exist then neither would testing. ISO 31000 – *Risk Management* provides the standard approach to risk management.

There are two main risk management activities:

1. risk analysis – the identification and assessment of risks;
2. risk control – the mitigation and monitoring of risks.

When test activities are selected, prioritised and managed using risk analysis and risk control, this is called risk-based testing.

In a project, two different types of risks are usually managed: project and product.

Project risks

Project risks are concerned with the ability of the project to deliver on its objectives, essentially to deliver on time and in budget. They are therefore related to the management and control of the project, and are usually managed by the project manager. They include:

- Organisational issues:
  - delays in delivery of the work products (e.g. code, documents, test environments);
  - inaccurate estimates;

179

- cost-cutting;
- late changes.

• People issues:

- skills and training or staff may be inadequate;
- staff shortages;
- personal issues between staff impacting progress;
- users, business staff or subject-matter experts may be unavailable when needed;
- communication problems between project members and between the project and its stakeholders.

• Technical issues:

- weakness in the development process that impacts the quality of the work products;
- scope-creep (i.e. changes to project scope that could severely impact delivery dates);
- poor tool support.

• Supplier issues:

- failure of a third party to deliver on time or at all;
- financial issues with the supplier (e.g. bankruptcy);
- contractual issues, such as meeting acceptance criteria.

Product risks

Product risks are risks to the quality of the product. In other words, the potential of a defect occurring in the live environment is a product risk. Examples of product risks are:

- Failure-prone software delivered – not able to perform as intended according to the specification or the user requirements. For example, missing or wrong functionality, incorrect calculations, inefficient algorithms and runtime errors.
- System architecture may not adequately support non-functional requirements (e.g. security vulnerabilities, reliability and system performance).
- Usability issues for end users, such as a poor look and feel of the product and lack of help documentation and guidance.

When product risks occur in production systems there can be significant consequences, including:

- users stating that the product does not meet their needs and expectations;
- loss of revenue, trust and reputation for the business;
- damage to third parties;

- high maintenance costs;
- helpdesk overloaded with user queries and complaints;
- criminal penalties and legal liabilities;
- in extreme cases, physical damage, injuries or even death.

Product risk analysis

As referenced at the start of this section, if there was not a risk of defects or failures within the software and hardware development processes there would be no need for software testing. The risks that form the approach to risk-based testing are product risks.

Product risk analysis should occur at the start of the SDLC. This includes risk identification and risk assessment. **risk identification** generates a comprehensive list of the risks, often collated via the use of workshops, interviews, brainstorming and cause and effect diagrams. The stakeholders involved will be identified based on their understanding of the application/hardware being developed, with the use of historic defect logs related to the current development and previous risk analyses on earlier releases of the software or hardware.

Once the risks have been identified **risk assessment** takes place, starting with categorising similar risks into groups. For each risk or risk category found, a probability (chance of the risk being realised) and impact (what will happen if the risk is realised) should be identified, as well as the identification and management of any mitigating actions (actions aimed at reducing the probability of a risk occurring, or reducing the impact of the risk if it did occur). Categorisation helps with the identification of mitigating actions for all risks within a category, as risks that fall into the same category are likely to be mitigated using a similar approach.

Risk assessment can use a qualitative approach, a quantitative approach or a mixture of both. A qualitative approach to risk analysis can be achieved by using a risk matrix such as the one shown in **Figure 5.4**, based on high-level risk assessment for an upgrade to a mobile banking app.

In the example shown in **Figure 5.4**, risks are judged by their likelihood and impact and plotted on the matrix. Anything in the darkest section has high likelihood and high impact, the medium-shaded section is medium likelihood and impact and the lightest is low likelihood and impact.

Using the same example, a quantitative approach would be to allocate a number to the likelihood and impact. The login function failing would get a 4 for likelihood and a 3 for impact; multiply them together and you get 12. The higher the number, the higher the risk.

When it is possible to identify the actual specific financial loss if a risk materialises, then a slightly amended quantitative approach could be taken. If you knew that the risk probability was 3 (possible) so 50 per cent, and the actual impact would be £100,000, you could multiply the two together to get a fiscal view of the risk: 0.5 × £100,000 = £50,000 financial risk. The risk with the highest monetary value would become your

Figure 5.4 A high-level risk assessment

Impact →

	1 Negligible	2 Minor	3 Moderate	4 Major	5 Catastrophic
5 Very likely				Unable to transfer cash	
4 Probable			Login function fails	If system stops mid-transaction the transaction is lost	Total app lock out
3 Possible			App performance	Third party goes bankrupt	Security vulnerability enabling others to access user account
2 Not Likely			Branding colours not correct		
1 Very unlikely	Doesn't install on certain phones				Data leak of all user bank details

Likelihood

priority risk to mitigate. This model works exceptionally well when the project is fully aware of the costs of failure; where that is not the case and it is not possible to get an agreement on the cost of a risk materialising, then using the 1–5 numeric approach will work best.

Product risk analysis can influence the structure and thoroughness of the testing. It can:

- Determine the scope of testing to be carried out.
- Determine the test levels and propose test types to be performed.
- Determine the test techniques to be employed and the coverage to be achieved. For example, three-value boundary value analysis (BVA) may be chosen over two-value BVA and branch coverage chosen over statement coverage for the highest risk components under test.
- Estimate the test effort required for each task.
- Prioritise testing to find the critical defects as early as possible.
- Determine whether any activities in addition to testing could be employed to reduce risk.

Product risk control

During product risk control, all of the measures that are taken in response to the identified and assessed risks are reviewed using risk mitigation and risk control approaches.

During the risk assessment, suggestions on the actions required to mitigate the risk are made to reduce the risk level – for example, a possible mitigating action against the risk of poor requirements could be using formal reviews as soon as the requirements have been documented at the start of a project. During product risk control, **risk mitigation** implements the identified mitigating actions, whereas **risk monitoring** is there to ensure the mitigating actions are effective at reducing the level of risk, to review and collect information to improve the risk assessment and to identify any new risks.

Product risks also provide information enabling decisions regarding how much testing should be carried out on specific components or systems – for example, the more risk there is, the more detailed and comprehensive the testing may be. In these ways testing is used to reduce the risk of an adverse effect (defect) occurring or being missed.

When reviewing product risks there are several possible risk outcomes to be considered. These include risk mitigation through testing, risk acceptance (acceptance that the risk can't be mitigated), risk transfer (allocating the risk to someone better placed to mitigate it) or a contingency plan of what to do when a risk occurs.

If testing is selected as the mitigating action, this can take many forms, including:

- Select the testers with the right level of experience and skills, suitable for a given risk type.
- Apply an appropriate level of independence of testing.
- Conduct reviews and perform static analysis.
- Apply the appropriate test techniques and coverage levels.
- Apply the appropriate test types addressing the affected quality characteristics.
- Perform dynamic testing, including regression testing.
- Perform differing levels of coverage of dynamic testing; the higher the risk, the higher the coverage, etc.

Mitigating product risks may also involve non-test activities. For example, in the poor requirements situation, a better and more efficient solution may be simply to replace (or train) the analyst who is writing the poor requirements in the first place.

Risk-based testing draws on the collective knowledge and insights of the project stakeholders, testers, designers, technical architects, business reps and anyone with knowledge of the solution to determine the risks and the levels of testing required to address those risks.

Testing is a risk control activity that provides feedback about the residual risk in the product by measuring the effectiveness of critical defect removal and by reviewing the effectiveness of contingency plans.

CHECK OF UNDERSTANDING

1. What are the two types of risks that must be considered in testing?
2. Compare and contrast these two risk types.
3. How early in the life cycle can risk impact the testing approach?
4. What are the activities that can be deployed to reduce a risk to be mitigated by testing?

TEST MONITORING, TEST CONTROL AND TEST COMPLETION

Having developed the test plan, the activities and timescales determined within the test execution schedule need to be constantly reviewed against what is happening. This is **test monitoring**. The purpose of test monitoring is to provide feedback and visibility of the test progress, and whether the exit criteria and the tasks associated with the exit criteria will be satisfied. These could include meeting the target for the mitigation of product risks, or requirements coverage, defect leakage or acceptance criteria.

The data required to monitor progress can be collected manually; for example, counting test cases developed at the end of each day. Alternatively, using a test management tool to collect the data as an automatic output from the tool, either already formatted into a report or as a data file that can be manipulated to present a picture of progress.

Having implemented test monitoring to understand progress through the test plan, test control is the corrective action undertaken for issues identified through test monitoring. Slippage of test activity dates or delays in delivery of external components are two potential issue areas. Test-control actions could include:

- reprioritising tests when an identified risk becomes an issue;
- changing the test schedule to address delays – for example, in the delivery of a test environment or the code;
- re-evaluating whether a test item meets entry/exit criteria due to rework;
- adding new resources when and where needed;
- changing the scope of the test activity.

The following test-control activities are likely to be outside the test manager's responsibility. However, this should not stop the test manager making a recommendation to the project manager, such as:

- Descoping of functionality – that is, removing some less important planned deliverables from the initial delivered solution to reduce the time and effort required to achieve that solution.
- Delaying release into the production environment until exit criteria have been met.

- Continuing testing after delivery into the production environment so that defects are found before they occur in production. (This requires knowledge of when delivered functionality is most likely to be used in the live system. For example, continuing the testing of an end-of-quarter financial reporting function that will not run for two months after the live release date.)

Test completion, as discussed in **Chapter 1**, is the collection of data from completed test activities to consolidate test experience, any testware and any other relevant information. By their nature, test completion activities formally occur at project milestones such as when a test level is completed, an Agile iteration is finished, a test project is completed or cancelled, a software system is released or a maintenance release is completed. This collected data is fed into a test completion report – see the section 'Purpose, content and audience for test reports'.

Metrics used in testing

Test metrics are gathered to show progress against the planned schedule, the current quality level of the test object and the effectiveness of the test activity in achieving the iteration goals. In any project, test metrics can be collected at any time – either during or at the end of the project – to assess:

- project progress (e.g. task completion, resource usage, test effort);
- test progress (e.g. test case implementation progress, test environment preparation progress, number of test cases run/not run, passed/failed, test execution time);
- product quality (e.g. availability, response time, mean time to failure);
- defects (e.g. number and priorities of defects found/fixed, defect density, defect detection percentage);
- risk (e.g. residual risk level);
- coverage (e.g. requirements coverage, code coverage);
- cost (e.g. cost of testing, organisational cost of quality).

Ultimately, test metrics are used to track progress towards the completion of testing or iteration, which is determined by the exit criteria or Definition of Done.

Purpose, content and audience for test reports

Test reporting is the process whereby test metrics are reported both during and at the end of a test activity, to update the stakeholders regarding the testing tasks undertaken. Test reports produced during the test activity are referred to as **test progress reports**, whereas a test report produced after a test activity has completed may be referred to as a **test completion report**.

The test team regularly (daily, weekly and monthly) issues a test progress report during test planning and execution for the project stakeholders, such as the sponsor, project and programme managers and any product owners. When exit criteria have been met and a test activity completes, the test manager issues a test completion report. This report provides an overview of the test activity undertaken, using data derived from the test progress reports.

The following information may be included in a test progress report:

- test period;
- test progress (e.g. ahead or behind schedule), including any notable deviations;
- impediments for testing, and their workarounds;
- metrics collected during testing (see the 'Test metrics' section for examples);
- new and changed risks within the testing period;
- testing planned for the next period.

In addition to the above, test progress reports may also include:

- other factors impacting progress;
- the quality of the test object.

A test completion report will be produced when the test activity has been completed, when a project, test level or test type is complete and either the exit criteria are met or it is agreed that sufficient exit criteria have been met to complete testing.

The content of a typical test completion report includes:

- test summary;
- testing and product quality evaluation based on the original test plan (i.e. test objectives and exit criteria);
- deviations from the test plan (e.g. differences from the planned schedule, duration and effort);
- testing impediments and workarounds;
- test metrics based on test progress reports;
- unmitigated risks, defects not fixed;
- lessons learned that are relevant to the testing.

ISO/IEC/IEEE 29119-3 documents required contents for both test progress reports (called test status report in the standard) and a test completion report.

Tables 5.1a and **5.1b** detail the two separate standard contents.

Reporting of progress within a team is often frequent and informal, but can be formal; however, test completion follows a set template and happens only once, at the end of the test activity.

The information gathered can also be used to help with any process improvement opportunities. This information can be used to assess whether:

- the goals for testing were correctly set (were they achievable; if not why not?);
- the test approach or strategy was adequate (e.g. did it ensure there was enough coverage?);
- the testing was effective in ensuring that the objectives of testing were met.

Table 5.1a Test progress report outline

Section no.	Heading	Details
1	Overview	Identifies the document and describes the origins and history
2	Unique identification of the document	The specific unique identifier allocated to this document, e.g. TP 00001
3	Issuing organisation	Specifies who is responsible for completion and distribution of the document
4	Approval authority	Identifies who is responsible for reviewing and signing off the document before it is issued
5	Change history	A record of each version of the document and any changes that were included for each version
6	Introduction	Explanatory information about the content and structure of the document
7	Scope	Defines the areas of coverage included within the document, test activities, etc.
8	References	Lists referenced documents and identifies repositories for system, software and test information. The references may be separated into 'external' references that are imposed from outside the organisation and 'internal' references that are imposed from within the organisation
9	Glossary	A glossary that defines the terms, abbreviations and acronyms, if any, used in the document
10	Test status	Includes: • reporting period • progress against the test plan • factors blocking progress • test measures • new and changed test risks • testing planned in the next period

Table 5.1b Test completion report outline

Section no.	Heading	Details
1	Overview	Identifies the document and describes the origins and history
2	Unique identification of the document	The specific unique identifier allocated to this document, e.g. TSR 00001
3	Issuing organisation	Specifies who is responsible for completion and distribution of the document
4	Approval authority	Identifies who is responsible for reviewing and signing off the document before it is issued
5	Change history	A record of each version of the document and any changes that were included for each version
6	Introduction	Explanatory information about the content and structure of the document
7	Scope	Defines the areas of coverage included within the document, test activities, etc.
8	References	Lists referenced documents and identifies repositories for system, software and test information. The references may be separated into 'external' references that are imposed from outside the organisation and 'internal' references that are imposed from within the organisation
9	Glossary	A glossary that defines the terms, abbreviations and acronyms, if any, used in the document
10	Testing performed	Includes: • a summary of testing performed • deviations from planned testing • test completion evaluation, e.g. have exit criteria all been met; if not why not? • factors that blocked progress • the collated test measures • residual risks • test deliverables • reusable test assets • lessons learned

Communicating the status of testing

Key for any report is that it is focused on the information required based upon the report's audience. Stakeholders will have different requirements; for example, some recipients may require graphs, while others wish to see the detailed data. The level of reporting may depend on test management concerns, organisational test strategies, regulatory standards or, in the case of self-organising teams, the team itself. It is the responsibility of the test team to ensure that the recipients' requirements for reports are understood before any test activity starts.

In an Agile project, the test progress reporting may be included in task boards and burn-down charts, which may also be discussed in the daily stand-up meeting, where the project team reviews progress during the previous period (often a day) and what is planned for the next period.

The best means of communicating test status varies. The options include:

- verbal communication with team members and other stakeholders;
- dashboards (e.g. continuous integration and continuous delivery/deployment (CI/CD) dashboards, task boards and burn-down charts);
- electronic communication channels (e.g. email, chat);
- online documentation;
- formal test reports.

One or more of these options may be used. Where teams are distributed and face to face isn't possible, it is more appropriate to provide a more formal communication method.

CHECK OF UNDERSTANDING

1. Name four common test metrics.
2. List three topics that appear in a test completion report that don't appear in a test progress report.
3. Identify three ways a test manager can control testing if there are more tests than there is time to complete them.

CONFIGURATION MANAGEMENT

The purpose of configuration management is to establish and maintain the integrity of the component or system, the testware, and their relationships to one another throughout the project and product life cycle. It involves managing products and processes by managing the information about them, including changes, and ensuring that they are what they are supposed to be in every case. Anything managed via configuration management is called a configuration item.

For testing, configuration management will involve identifying, controlling and tracking work products used throughout the SDLC; for example, test plans, test strategies, test conditions, test cases, test scripts, test results, test logs and test reports are configuration items.

Configuration management should ensure that each test item is uniquely identified and provide full traceability throughout the test process; for example, a requirement should be traceable through to the test cases that are produced to test it, and vice versa.

Effective configuration management is important for the test process, as the contents of each release of software into a test environment must be understood and be the correct version. Otherwise testers could end up wasting time because either they are testing an invalid release of the software or the release does not integrate successfully, leading to the failure of many tests.

For a complex configuration item such as a test environment, configuration management records the items it consists of, their relationships and versions. If the configuration item is approved for testing, it becomes a baseline and can only be changed through a formal change control process.

When a baseline changes, configuration management keeps track of what changes are made, the version number and when they were made.

In most instances the project will have already established configuration management processes that will define the documents and code to be held under configuration management. If this is not the case, then during test planning the process and tools required to establish the right configuration management processes will need to be selected/implemented by the test manager.

It is important that configuration management supports testing, by ensuring the following:

- All configuration items, including test items (individual parts of the test object), are uniquely identified, version controlled, tracked for changes and related to other configuration items so that traceability can be maintained throughout the test process.
- All identified documentation and software items are referenced unambiguously in test documentation.

Continuous integration, continuous delivery, continuous deployment and the associated testing are typically implemented in projects as part of an automated DevOps pipeline (see **Chapter 2**), in which automated configuration management is normally included.

CHECK OF UNDERSTANDING

1. Define configuration management.
2. What can be stored under configuration management?
3. Why is it important to have effective configuration management?

DEFECT MANAGEMENT

A defect is any unplanned event occurring that requires further investigation. In testing, this translates into anything where the actual result is different from the expected result. A defect, when investigated, may be a true defect; however, it may also be a change to a specification or an issue with the test being run. It is important that a process exists to track all defects through to closure. This process must be agreed by all parties involved and can be quite informal.

Defects can be raised at any time throughout the SDLC, including reviews of the test basis (requirements, specifications, etc.), coding, static analysis, test specification and dynamic testing.

Typical defect reports have the following objectives:

- To provide developers and other parties responsible for handling and resolving defects with feedback on the problem to enable identification, isolation and correction as necessary. It must be remembered that most developers and other parties who will correct the defect or clear up any confusion will not be present at the point of identification, so without full and concise information they will be unable to understand the problem, and possibly therefore be unable to understand how to go about fixing it. The more information provided, the better.

- To provide a means of tracking the quality of the system under test and the progress of the testing. Key metrics used to measure progress is a view of how many defects are raised, their priority and finally that they have been corrected and signed off.

- To provide ideas for test process improvement. For each defect the point of injection should be documented – for example, a defect in requirements or code – and by applying root cause analysis (discussed in **Chapter 1**) subsequent process improvement can focus on that area to stop the same defect occurring again.

A defect report filed during dynamic testing typically includes:

- unique identifier;
- title with a short summary of the anomaly being reported;
- date when the anomaly was observed, issuing organisation and author, including their role;
- identification of the test object and test environment;
- context of the defect (e.g. test case being run, test activity being performed, SDLC phase and other relevant information such as the test technique, checklist or test data being used);
- description of the failure to enable reproduction and resolution, including the steps that detected the anomaly, and any relevant test logs, database dumps, screenshots or recordings;
- expected results and actual results;

- severity of the defect (degree of impact) on the interests of stakeholders or requirements;

- priority to fix;

- status of the defect (e.g. open, deferred, duplicate, waiting to be fixed, awaiting confirmation testing, reopened, closed, rejected);

- references (e.g. to the test case).

Defect management is the process of recognising, investigating, taking action and disposing of defects. It involves recording defects, classifying them and identifying the impact. The process of defect management ensures that defects are tracked from recognition to correction, and finally through retest and closure. It is important that organisations document their defect management process and ensure that they have appointed someone (often called a defect manager/coordinator) to manage/police the process.

Defects are raised on defect reports, either electronically via a defect management system (from Microsoft Excel to sophisticated defect management tools) or on paper. Some of this data may be automatically included when using defect management tools (e.g. identifier, date, author and initial status).

The syllabus also recognises that ISO/IEC/IEEE 29119-3 defines a test defect report (called an incident report), which has sections aligned with those documented above.

To assist in your defect management learning we have included an exercise below. The answer can be found at the end of this chapter:

EXERCISE 5.1

Defect id	138	
Date Issued	06/09/2023	
Title	Cancelling from Booking page logs out rather than return to home page	
Description	When I pressed 'Cancel' on the main booking screen I was expecting to return to the customer's Home page but instead the system logged me out, so I had to log in again. User Story B-37 states that Cancel should return the customer to their Home page.	
Severity	High	
Status	New	
Expected Results	After selecting to Cancel the booking, customer is returned to their Home page.	
Actual Results	Customer is logged-out.	

The above defect report shows the template currently in use for a new online booking system. The retrospective meeting for the first iteration highlighted several issues with defect management. The key issues were:

- There has been a high incidence of defect reports being returned by the development team, asking for more information to enable analysis and reproduction.
- Many defect reports were being returned to the wrong originator.
- Some defect reports that were fixed by the development team passed retest by the testers, but these later turned out to be 'false negatives' – that is, the fixes had been incorrectly applied but went undetected.

Which defect report fields, described in this chapter, should be added to the template for the next iteration, to address these issues?

CHECK OF UNDERSTANDING

1. Identify three details that are usually included in a defect report.
2. What is the name of the standard that includes an outline of a test defect report?
3. What is a test defect?

SUMMARY

In this chapter we have looked at the component parts of test management. We started with test planning and understanding the purpose of test planning being to describe the objectives, resources and processes for a test project. We referred to ISO/IEC/IEEE 29119-3, which provides outlines of four test planning documents:

- the test plan;
- the test progress report;
- the test completion report;
- the test defect report.

We looked at the contribution made by testers in iteration and release planning, focusing on the key role a tester plays in refining user stories and defining and redefining the product backlog, as well as their involvement in release and iteration planning, including risk analysis and estimating effort.

Test management depends not only on the preparation of the required documents but also on the development of the right entry and exit criteria and estimates, the monitoring of progress through the plan and the control activities implemented to ensure the plan is achieved.

We then looked at test case prioritisation and the three techniques widely used when prioritising the test activities.

The test pyramid is a model used to show that tests developed for different levels of testing may have different granularity. Three layers of tests are commonly used in this model, but you can have as many levels as needed when in actual use.

Next we looked at Marick's testing quadrant and how it defines the test levels and test types to be used in Agile development, as well as providing a useful way of presenting the testing undertaken in an Agile project.

We explored risk and testing. When developing the test plan, the test team will look at the product risks (risks that relate directly to the failure of the product in the live environment) to decide the level of testing required, as well as ensuring that any project risks (risks relating to the delivery of the project) are mitigated. The risks will initially be identified and assessed (product risk analysis), and then through mitigating actions risks will be controlled (product risk control).

After a plan of activity has been developed and time begins to pass, progress of the test items needs to be monitored. If any activity is delayed or there has been a change of any kind in the project itself, test control may need to be applied – for example, the test plan may need to be revised or other actions may be required to ensure that the project is delivered on time. Finally, we looked at test completion as the activity to collect all relevant data undertaken at specific milestones, such as the completion of an Agile iteration.

As part of the monitoring process, we looked at the kind of metrics used in testing, such as project progress metrics and risk metrics, and the two main test reports: the test progress report and the test completion report. We also looked at how test status is reported (e.g. verbally or via a dashboard).

We then looked at configuration management. When running test cases against the code, it is important that the tester is aware of the version of code being tested and the version of the test being run. Controlling the versioning of the software and test assets is called configuration management. Lack of configuration management may lead to issues like loss of already-delivered functionality, reappearance of previously corrected defects and no understanding of which version of the test was run against which version of the code.

Finally, we explored how the defects found during testing are recorded, and we reviewed the level of detail that needs to be recorded to ensure that any defect is fully understood and that any fix then made is the right one.

Example examination questions with answers below

E1. K2 question
Which of the following correctly identify a metrics-based approach to estimation?

    A.   Groups of experts provide estimates based on their experience.

    B.   Obtaining two estimates, most optimistic and most pessimistic.

    C.   Records of defects found in a similar stage in another project and the time taken to remove them.

    D.   Comparison of the estimates given by testers on the project and independent experts.

E2. K2 question
Which of the following is appropriate content for a test summary report?

    i.   The status of testing and progress against the test plan.

    ii.   Information about what occurred during a test period.

    iii.   A review of test activity progress and resource consumption for the system testing phase.

    iv.   An assessment of the quality of the test object at the present stage of testing.

    A.   i and ii.

    B.   ii and iii.

    C.   iii and iv.

    D.   i and iv.

E3. K1 question
Which of the following terms is used to describe the management of software components comprising an integrated system?

    A.   Configuration management.

    B.   Defect management.

    C.   Test monitoring.

    D.   Risk management.

E4. K1 question
A new system is about to be developed. Which of the following functions has the highest level of risk?

    A.   Likelihood of failure = 20 per cent; impact value = £100,000.

    B.   Likelihood of failure = 10 per cent ; impact value = £150,000.

    C.   Likelihood of failure = 1 per cent ; impact value = £500,000.

    D.   Likelihood of failure = 2 per cent ; impact value = £200,000.

E5. K2 question
Which of the following statements about risks is most accurate?

    A. Project risks rarely affect product risk.

    B. Product risks rarely affect project risk.

    C. A risk-based approach is more likely to be used to mitigate product rather than project risks.

    D. A risk-based approach is more likely to be used to mitigate project rather than product risks.

Answers to the self-assessment questions in the chapter

SA1. The correct answer is C.

SA2. The correct answer is A.

SA3. The correct answer is A.

Answer to the defect management exercise

- There has been a high incidence of defect reports being returned by the development team, asking for more information to enable analysis and reproduction.
  - *Identification of the test object and test environment.* Essential information for the developers to try to reproduce the problem. Perhaps there are issues with the test environment?
- Many defect reports were being returned to the wrong originator.
  - *Author. The developers need to know exactly who to direct their queries to.*
- Some defect reports that were fixed by the development team passed retest by the testers, but these later turned out to be 'false negatives'.
  - *References (e.g. the test case). It seems likely that the retests are not following the exact steps taken when the original failure occurred. The defect reports are limited in respect to this information (in the example, does the failure only occur with certain types of customer?). It is important that the test case that identified the failure is run again during retest, in the same way to avoid defect masking.*

Answers to example examination questions

E1. The correct answer is C.

- A. This approach is known as the Wideband Delphi estimation technique and is based on multiple, well-informed estimates but not on data, as required by a metrics-based approach.
- B. Most optimistic and most pessimistic estimates is an expert-based approach used in three-point estimation.
- C. This is a valid metrics-based approach because data about the volume of defects and the time taken to remove them is provided, albeit on a different project.
- D. This is potentially an effective way to improve the ability of testers to estimate accurately, but it is expert-based rather than metrics-based because there is no actual data to provide measures of achievement.

E2. The correct answer is B.

While all of the options describe content that summarises activity, option i is focused on identifying specific progress information – that is, whether or not progress is consistent with the plan. Item iv is also specific to progress towards an objective (the quality of the test object). Items ii and iii, in contrast, provide a broader view of what happened and what has been done – that is, they summarise rather than report specific progress. For these reasons, items ii and iii are more appropriate to a test summary report, while items i and iv are more appropriate to a test progress report.

E3. The correct answer is A.

Defect management is the collection and processing of defects raised when errors and defects are discovered. Test monitoring identifies the status of the testing activity on a continual basis. Risk management identifies, analyses and mitigates risks to the project and the product. Configuration management is concerned with the management of changes to software components and their associated documentation and testware.

E4. The correct answer is A.

In option B the product of probability × impact has the value £15,000; in option C the value is £5,000; and in option D it is £4,000. The value of £20,000 in A is therefore the highest.

E5. The correct answer is C.

In general, project risk and product risk can be hard to differentiate. Anything that impacts on the quality of the delivered system is likely to lead to delays or increased costs as the problem is tackled. Anything causing delays to the project is likely to threaten the delivered system's quality. The risk-based approach is an approach to managing product risk through testing, so it impacts most directly on product risk.

6 TEST TOOLS

Peter Williams

INTRODUCTION

As seen in earlier chapters, there are many tasks and activities that need to be performed during the test process. In addition, other tasks need to be performed to support the test process.

In order to assist in making the test process easier to perform and manage, many different types of test tools have been developed and used for a wide variety of testing tasks. Some of them have been developed in-house by an organisation's own software development or testing department. Others have been developed by software houses (also known as test-tool vendors) to sell to organisations that perform testing. More recently, open source tools have been developed that can be reused and enhanced. Even within the same type of tool, some will be home-grown while others will be developed as open source tools or by test-tool vendors.

This chapter describes the most commonly used types of tools that support the test process as well as the potential benefits and pitfalls associated with test automation.

Learning objectives

The learning objectives for this chapter are listed below. You can confirm that you have achieved these by using the self-assessment questions immediately following the learning objectives, the 'Check of understanding' boxes distributed throughout the text and the example examination questions provided at the end of the chapter. The chapter summary will remind you of the key ideas.

Each learning objective is allocated a K number to represent the level of understanding required; see the **Introduction** (pp. 2–3) for an explanation of K numbers.

Tool support for testing

- FL-6.1.1 (K2) Explain how different types of test tools support testing.

Benefits and risks of test automation

- FL-6.2.1 (K1) Recall the benefits and risks of test automation.

Self-assessment questions

The following questions have been designed to enable you to check your current level of understanding for the topics in this chapter. The answers are at the end of the chapter.

Question SA1 (K2)
Which of the following pairs of test tools are *likely* to be *most useful* during the test analysis stage of the test process?

    i.   Test execution tool.
    ii. Test data preparation tool.
    iii. Test management tool.
    iv. Requirements management tool.

    A. i and ii.
    B. i and iv.
    C. ii and iii.
    D. iii and iv.

Question SA2 (K1)
Which of the following is *most likely* to be a risk of using a test automation tool?

    A. Underestimating the demand for a tool.
    B. The purchase price of the tool.
    C. Overreliance on the tool.
    D. The cost of resources to implement and maintain the tool.

Question SA3 (K2)
What benefits do static analysis tools have over test execution tools?

    A. Static analysis tools find defects earlier in the life cycle.
    B. Static analysis tools can be used before code is written.
    C. Static analysis tools test that the delivered code meets business requirements.
    D. Static analysis tools are particularly effective for regression testing.

WHAT IS A TEST TOOL?

Definition of a test tool

A test tool can be thought of as a piece of software that is used to make the test process more effective or efficient. In other words, anything that makes testing easier, quicker, more accurate and so on.

This book focuses on those test tools listed in the 2023 syllabus and which are, generally, the test tools that are most commonly used in the test process.

Let us consider the building of a new hotel and examine the similarities with the introduction and use of test tools. Test tools need to be thought of as long-term investments that need maintenance to provide long-term benefits. Similarly, building a hotel requires a lot of upfront planning, effort and investment. Even when the hotel is ready for use, there is still a continual long-term requirement for the provision of services such as catering, cleaning, building maintenance, provision of staff to provide ad hoc services to customers and so on. From time to time there is a need for upgrades to infrastructure to keep up with new technology and customer demands. The long-term benefit is that this upfront investment and ongoing maintenance and support can provide substantial income in return.

In addition, there are risks that, over a period of time, the location of the hotel will become less attractive, resulting in lower demand, lower usage and a maintenance cost that is greater than the income received. Therefore, the initial investment is wasted because the ongoing need/requirement does not exist.

The graph in **Figure 6.1** demonstrates a typical payback model for implementing a test automation tool. The same principle applies to the majority of test tools. Note that there is an ongoing maintenance cost of using the tool, but this ongoing maintenance cost needs to be less than the cost of performing testing activities without the tool if the investment is to be worthwhile.

Figure 6.1 Test tool payback model

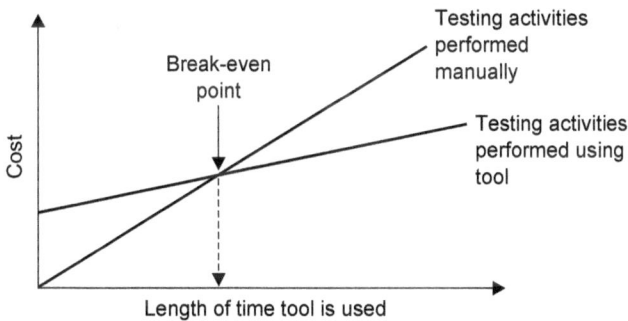

We discuss the benefits and risks of test automation later in this chapter.

Other tools can be built by developers in-house as the need arises. For instance, test harnesses, test oracles or test data preparation tools may be relatively easy to produce for developers with a good understanding of the tool requirements and the systems and databases in the test environment. More recently, open source tools have allowed developers and testers to use tools as the building blocks for developing in-house tools to meet specific needs. In addition, test tools have been developed by the UK Financial Conduct Authority for use by banks and building societies that participate in the Faster Payments and Account Switcher schemes.

EXAMPLE: HOTEL CHAIN SCENARIO

An example of a hotel chain with several UK-based hotels will be used throughout this chapter. The systems that comprise the organisation's system architecture are shown in **Figure 6.2**.

The general public can book rooms at any of the chain's hotels by:

- contacting staff in the hotel, who then use a graphical user interface (GUI) front-end to make the booking;
- telephoning customer services, who then use a GUI front-end to make the booking;
- using the company's website to make the booking online;
- using a mobile app that can be downloaded from the hotel chain's website.

Figure 6.2 Hotel system architecture

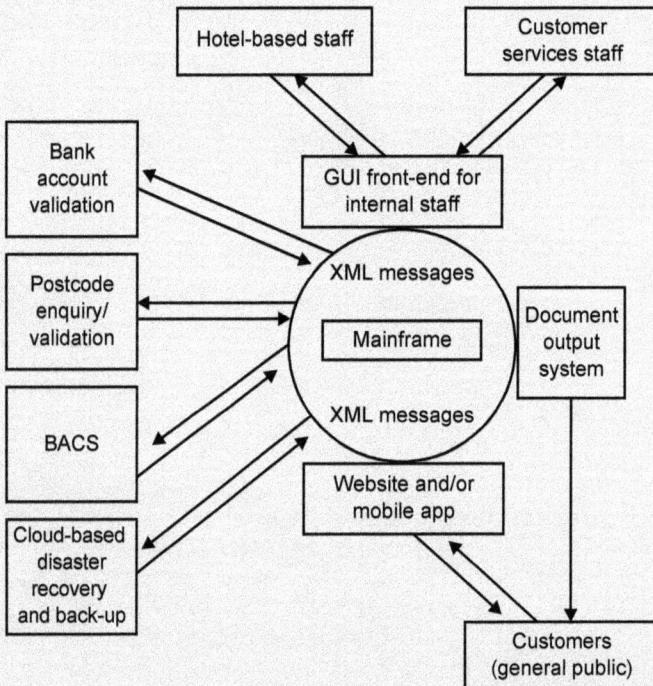

In all cases, communication with the mainframe computer is done via a middleware layer of Extensible Markup Language (XML) messages.

There is a document production system that produces PDF versions of customer correspondence such as booking confirmations, bills, invoices and so on. These are stored securely and can be downloaded by customers from the website.

Direct debit payments are made via Bankers Automated Clearing Services (BACS). Files are transmitted and confirmation and error messages are received back. Credit card payments can also be made. An enhancement to the security systems is being made to comply with Payment Card Industry standards. Payments can also be made by leading electronic payment systems (e.g. PayPal).

Validation of bank account details is performed by sending XML messages to and from a third-party system.

Validation and enquiry of address and postcode is also performed by sending XML messages to and from a third-party system.

A new release of the system is planned for six months' time. This will include:

- Code changes to replace the XML middleware layer. Mainframe changes will be performed by an outsourced development team in India.
- Various changes to website screens to improve usability.
- The introduction of a new third-party calendar object from which dates can be selected.
- Removal of the ability for customers to pay by cheque.
- An amended customer bill, plus two other amended documents.
- Two new output documents.
- Fixes to various existing low and medium severity defects.
- Improvements to disaster recovery by using cloud-based methods.
- Ongoing enhancements to the mobile app using Agile development methods. These will be deployed to production approximately every three weeks.

CHECK OF UNDERSTANDING

1. Would you expect a quick return on your investment in test tools? Why?
2. What is the advantage of developing a tool 'in-house'?

TOOL SUPPORT FOR TESTING

Types of tool

There are several ways in which test tools can be classified. They can be categorised according to:

- their purpose;
- the test process and the software development life cycle (SDLC) with which they are primarily associated;
- the type of testing that they support;
- the source of tool (shareware, open source, free or commercial);
- the technology used;
- who uses them.

In this book, test tools will be classified according to the type of activity they support.

Management tools
Test management tools and application life cycle management tools Test management tools and application life cycle management (ALM) tools provide support for various activities and tasks throughout the test process. The main difference between a test management tool and an ALM tool is that:

- A test management tool tends to focus on the test process and allow integration with other tools (but this integration may need to be built using APIs).
- An ALM tool typically has built-in integration with other tools.

In the remainder of this section any service/function provided by a test management tool will also be met by an ALM tool unless otherwise stated.

Standalone test management tools tend to be cheaper than ALM tools.

The diagram in **Figure 6.3** shows how a test management tool is the hub or centre of a set of integrated test tools.

Test management tools provide an architecture for creating, storing and editing test procedures. These may be linked or traced to requirements, test conditions and risks. Such test procedures can then be prioritised or grouped together and scheduled so that they are run in the most effective and efficient order. Some test management tools allow dependencies to be recorded so that tests that will fail owing to a known defect can be highlighted and left unexecuted. This allows testers to be redirected to run the highest priority tests available rather than waste their time and the test data they have prepared on tests that are certain to fail.

Tests can be recorded as passed or failed and usually a test management tool provides an interface to a defect management tool so that a defect can be raised if the actual and expected results differ.

Figure 6.3 An integrated set of tools

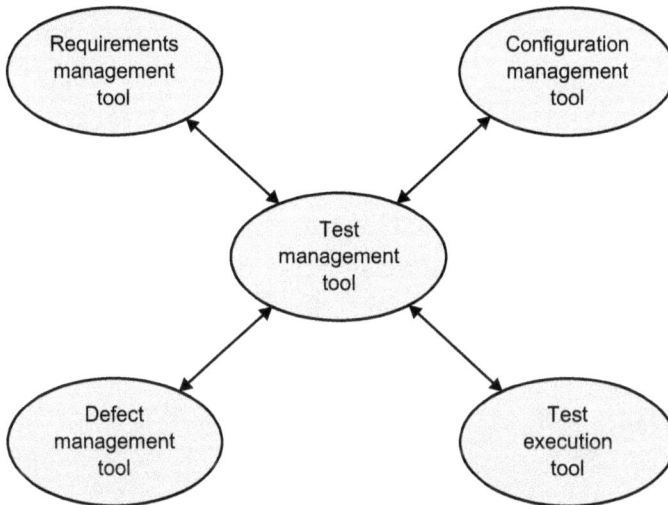

Test management tools can provide management information and reports on test procedures that are passed or failed. The amount of integration with other specialist tools is significant here. For instance, integration with requirements management tools allows reports to be produced on test progress against one or more requirements. Integration with defect management tools also allows reports to include analysis of defects against requirements.

Test management tools generally hold data in a database. This allows a large amount of reports and metrics to be produced. The metrics produced can be used as inputs to:

- test and project management to control the current project;
- estimates for future projects;
- identifying weaknesses or inefficiencies in the development or test processes that can be subsequently investigated with the aim of improving them.

Test management information reports should be designed to meet the needs of project managers and other key users. It may be necessary to export/import data in appropriate formats to other tools, such as:

- spreadsheets;
- project management/scheduling tools;
- management accounting systems;
- human resources/personnel systems and so on.

A test management tool can also enable reuse of existing testware in future test projects.

USE IN HOTEL CHAIN SCENARIO

In the scenario, a test management tool can be used to write down and store requirements for new functionality and subsequently to hold the test conditions necessary to test these requirements.

It can also be used to record whether tests have passed or failed and to produce test management information on progress to date.

Additionally, requirements and test conditions from previous developments will already exist in the test management tool. These can be used as the basis of any regression testing required. Indeed, a regression test pack may already exist. Clearly the regression test pack would have to be reviewed and amended as necessary to make it relevant to this release. However, the benefit is that much of the previous work could be reused, which, in turn, means that much less effort will be involved in creating a regression test pack.

Defect management tools   Defect management tools (also known as incident management tools) are one of the most widely used types of test tool. At a basic level, defect management tools are used to perform two critical activities: creation of a defect report; and maintenance of details about the defect as it progresses through the defect life cycle.

The level of detail to be captured about the defect can be varied, depending on the characteristics of the tool itself and the way in which the defect management tool is configured and used by the test organisation.

For example, the defect management tool can be configured so that lots of mandatory information is required in order to comply with industry or generic standards such as ISO/IEC/IEEE 29119-3. In addition, workflow rules may also be applied to ensure that the agreed defect life cycle is strictly adhered to, with defects only able to be assigned to particular teams or users. Alternatively, the tool can be configured to require very limited mandatory information, with most fields being free-format.

Defect management tools also use a database to store and manage details of defects. This allows the defect to be categorised according to the values stored in appropriate fields. Such values will change during the defect life cycle as the defect is analysed, debugged, fixed and retested. It is often possible to view the history of changes made to the defect.

The database structure also enables defects to be searched and analysed (using either filters or more complex Structured Query Language (SQL) type queries). This provides the basis for management information about defects. Note that as the values held against each defect change, the management information will also change. Therefore, users need to be aware of the danger of using outdated reports.

This data can be used in conjunction with data held in test management tools when planning and estimating for future projects. It can also be analysed to provide input to test process improvement projects.

Fields in the database structure normally include:

- priority (e.g. high, medium, low);
- severity (e.g. high, medium, low);
- assignee (the person to whom the defect is currently assigned, e.g. a developer for debugging, a tester to perform retesting);
- status in the defect life cycle (e.g. new, open, fixed, reopened, closed).

This allows management information to be produced from the defect management database about the number of high-priority defects with a status of open or reopened that are assigned to, say, Peter Morgan, compared with the number assigned to Geoff Thompson.

ALM tools typically include fully integrated defect management tools as part of their core product, while other defect management tools can be integrated with test management, requirements management and test execution tools. Such integration enables defects to be input and traced back to test cases and requirements.

USE IN HOTEL CHAIN SCENARIO

A defect management tool can be used to raise new defects and process them through the defect life cycle until resolved. It can also be used to check whether defects (or similar defects) have been raised before, especially for defects raised during regression testing.

A defect management tool could also be used to prioritise defects so that developers fix the most important ones first. It could also highlight clusters of defects. This may suggest that more detailed testing needs to be done on the areas of functionality where most defects are being found, as it is probable that further defects will be found as well.

Requirements management tools Requirements management tools are used by business analysts to record, manage and prioritise the requirements of a system. They can also be used to manage changes to requirements – something that can be a significant problem for testers, as test cases are designed and executed to establish whether the delivered system meets its requirements. Therefore, if requirements change after tests have been written, then test cases may also need to change. There is also a potential problem of changes not being communicated to all interested parties, thus testers could be using an old set of requirements while new ones are being issued to developers.

The use of a traceability function within a requirements tool (or integrated with an ALM or test management tool) enables links and references to be made between requirements, functions, test conditions, defects and other testware items. This means that as requirements change, it is easy to identify which other items may need to change.

Some requirements management tools can be integrated with test management tools, while ALM tools typically enable requirements to be input and related to test cases within the ALM tool.

Requirements management tools also enable requirements coverage metrics to be calculated easily, as traceability enables test cases to be mapped to requirements.

As can be seen, traceability can create a lot of maintenance work, but it does highlight those areas that are undergoing change.

USE IN HOTEL CHAIN SCENARIO

A change is required to three PDF documents that are stored securely on the website so that customers can log in and download them. The requirements are documented in the requirements management tool. Testers obtain the requirements from the tool and begin to devise test conditions and test cases.

A subsequent change means that further changes are made to the requirements. The testers should be made aware of the changes so that they can provide input to the impact analysis. Traceability within a requirements management tool will also highlight the test conditions affected by the changed requirement. The testers can review the change in requirements and then consider what changes need to be made to the test conditions and test cases.

Configuration management tools and continuous integration tools Configuration management tools are designed primarily for managing the versions of different software (and hardware) components that comprise a complete build of the system, and various complete builds of systems that exist for various software platforms over a period of time.

Continuous integration (CI) tools have been developed more recently and can be used in conjunction with configuration management tools to ensure that the correct versions of different programs are integrated into the daily build that is deployed into the test environment. This is particularly advantageous for Agile developments, where it is important to produce builds automatically and quickly.

A **build** is a development activity where a complete system is compiled and linked (typically daily) so that a consistent system is available at any time, including all the latest changes.

USE IN HOTEL CHAIN SCENARIO

Within the hotel booking system there will be many versions of subsystems due to the date at which the version was included in a build, or the operating system on which the version works and so on. Each version of a subsystem will have a unique version number and will comprise many different components (e.g. web services, program files, data files, etc.).

The configuration management tool maps the version number of each subsystem to the build (or release) number of the integrated system. As shown in **Table 6.1**, Build A (UNIX) and Build B (Windows Server 2016) might use the same version (v1.02) of the Payments In subsystem, but Build C might use version v1.04.

Table 6.1 Configuration traceability

Build for integrated system	Version of Payments In system	Credit card payment test procedure ID	Electronic payment test procedure ID
Build A	v1.02	TP123a	TP201
Build B	v1.02	TP123b	TP201
Build C	v1.04	TP123b	TP201

The same principle applies to testware with a different version number for a test procedure being used, depending on the version number of the build. For instance, test procedure TP123a might be used for Build A and TP123b might be used for Build B – even though both have the same purpose and even expected results. However, another test procedure, TP201, may be applicable to all builds.

A CI tool will support the Agile development methods being used for the mobile app so that deployments into the test environment can be done automatically and quickly.

The amount of benefit to be obtained from using configuration management tools and CI tools is largely dependent on:

* the complexity of the system architecture;
* the number and frequency of builds of the integrated system;
* how much choice (options) is available to customers (whether internal or external).

For example, a software house selling different versions of a product to many customers who run on a variety of operating systems is likely to find configuration management tools more useful than an internal development department working on a single operating system for a single customer. However, an internal development department using an Agile approach will find CI tools almost essential for managing frequent deployments (daily builds) into the test environment.

The use of configuration management tools allows traceability between testware and builds of an integrated system and versions of subsystems and modules. Traceability is useful for:

- identifying the correct version of test procedures to be used;
- determining which test procedures and other testware can be reused or need to be updated/maintained;
- assisting the debugging process so that a failure found when running a test procedure can be traced back to the appropriate version of a subsystem.

CHECK OF UNDERSTANDING

1. What is traceability?
2. Which tool is likely to be most closely integrated with a requirements management tool?
3. Which tool would you use to identify the version of the software component being tested?
4. Which tool would you use to produce a daily build?

Static testing tools

Tools that support reviews   Tools that support reviews (also known as review tools or review process support tools in previous versions of the syllabus) provide a framework for reviews or inspections. This can include:

- maintaining information about the review process, such as rules and checklists;
- the ability to record, communicate and retain review comments and defects;
- the ability to amend and reissue the deliverable under review while retaining a history or log of the changes made;
- traceability functions to enable changes to deliverables under review to highlight other deliverables that may be affected by the change;
- the use of web technology to provide access from any geographical location to this information.

Review tools can interface with configuration management tools to control the version numbers of a document under review.

If reviews and inspections are already performed effectively, then a review tool can be implemented fairly quickly and relatively cheaply. However, if such a tool is used as a means for imposing the use of reviews, then the training and implementation costs will be fairly high (as is the case for implementing a review process without such tools).

These tools support the review process, but management buy-in to reviews is necessary if benefits from them are to be obtained in the long run.

Review tools tend to be more beneficial for peer (or technical) reviews and inspections rather than walkthroughs and informal reviews.

USE IN HOTEL CHAIN SCENARIO

The hotel company could use a review tool to perform a review of a system specification written in the UK, so that offshore developers can be involved in the review process. In turn, the review of program code, written offshore, could also be performed using such a tool. This means that both the UK and offshore staff could be involved in both reviews, with no excuses for the right people not being available to attend.

Static analysis tools

Static analysis tools are most commonly associated with code analysis, but can also be used against other types of work products that have a structure, such as models or text with a formal syntax. There are many tools available on the market that can perform spelling and grammar checking on written documents – indeed, Microsoft's Editor is built into MS Word to analyse documents and offer suggestions for spelling, grammar and stylistic issues.

Static code analysers examine code before it is executed in order to identify defects as early as possible. They are used mainly by developers prior to unit testing. A static analysis tool can generate lots of error and warning messages about the code. Training may be required in order to interpret these messages and it may also be necessary to configure the tool to filter out particular types of warning messages that are not relevant. The use of static analysis tools on existing or amended code is likely to result in lots of messages concerning programming standards. A way of dealing with this situation should be considered during the selection and implementation process. For instance, it may be agreed that small changes to existing code should not use the static analysis tool, whereas medium to large changes to existing code should use the static analysis tool. A rewrite should be considered if the existing code is significantly non-compliant.

Static analysis tools can find defects that are hard to find during dynamic testing. They can also be used to enforce programming standards (including secure coding), to improve the understanding of the code and to calculate complexity and other metrics (see **Chapter 3**).

Some static analysis tools are integrated with dynamic analysis tools and coverage measurement tools. They are usually language-specific, so to test code written in C++ you need to have a version of a static analysis tool that is specific to C++.

Other static analysis tools come as part of programming languages or only work with particular development platforms. Note that debugging tools and compilers provide some basic functions of a static analysis tool, but they are generally not considered to be test tools and are excluded from the ISTQB syllabus.

The types of defect that can be found using a static analysis tool can include:

- Syntax errors (e.g. spelling or missing punctuation).

- Variance from programming standards (e.g. too difficult to maintain).

- Invalid code structures (e.g. missing ENDIF statements).

- The structure of the code means that some modules or sections of code may not be executed. Such unreachable code or invalid code dependencies may point to errors in code structure.

- Portability (e.g. code compiles on Windows but not on UNIX).

- Security vulnerabilities.

- Inconsistent interfaces between components (e.g. XML messages produced by component A are not of the correct format to be read by component B).

- References to variables that have a null value or variables declared but never used.

USE IN HOTEL CHAIN SCENARIO

Static analysis tools may be considered worthwhile for code being developed by offshore development teams who are not familiar with in-house coding standards. Such tools may also be considered beneficial for high-risk functions such as BACS and other external interfaces.

CHECK OF UNDERSTANDING

1. Which of the tools used for static testing is/are most likely to be used by developers rather than testers?

2. In which part of the test process are static analysis tools likely to be most useful?

Test design and implementation tools
Test design tools   Test design tools are used to support the generation and creation of test cases. In order for the tool to generate test cases, a test basis needs to be input and maintained. Therefore, many test design tools are integrated with other tools that already contain details of the test basis, such as:

- requirements management tools;

- static analysis tools;

- test management tools.

The level of automation can vary and depends on the characteristics of the tool itself and the way in which the test basis is recorded in the tool. For example, some tools allow specifications or requirements to be specified in a formal language. This can allow test cases with inputs and expected results to be generated. Other test design tools allow a GUI model of the test basis to be created and then allow tests to be generated from this model.

Some tools (sometimes known as test frames) merely generate a partly filled template from the requirement specification held in narrative form. The tester will then need to add to the template and copy and edit as necessary to create the test cases required.

Tests designed from database, object or state models held in modelling tools can be used to verify that the model has been built correctly and to derive some test cases. Tests derived can be very thorough and give high levels of coverage in certain areas.

Some static analysis tools integrate with tools that generate test cases from an analysis of the code. These can include test input values and expected results.

A test oracle (not in the syllabus) is a type of test design tool that automatically generates expected results. However, these are rarely available because they perform the same function as the software under test. Test oracles tend to be most useful for:

- replacement systems;
- safety-critical systems;
- migrations;
- regression testing.

USE IN HOTEL CHAIN SCENARIO

A test oracle could be built using a spreadsheet to support the testing of customers' bills. The tester can then input details for calculating bills, such as the total bill based on various transaction types, refunds, VAT and so on. The spreadsheet could then calculate the total bill amount and this should match the bill calculated by the system under test.

However, test design tools should only be part of the approach to test design. They need to be supplemented by other test cases designed using other techniques and the application of risk.

Test design tools could be used by the test organisation in the scenario, but the overhead to input the necessary data from the test basis may be too great to give any real overall benefit. However, if the test design tool can import requirements or other aspects of the test basis easily, then it may become worthwhile.

Test design tools tend to be more useful for safety-critical and other high-risk software, where coverage levels are higher and industry, defence or government standards need to be adhered to. Commercial software applications, such as the hotel system, do not usually require such high standards and therefore test design tools are of less benefit in such cases.

Test data preparation tools Test data preparation tools are used by testers and developers to manipulate data so that the environment is in the appropriate state for the test to be run. This can involve making changes to the field values in databases, data files and so on, and populating files with a spread of data (including depersonalised dates of birth, names and addresses, etc. to support data anonymity).

USE IN HOTEL CHAIN SCENARIO

A set of test data may be created by taking, say, 5 per cent of all records from the live system and scrambling personal details so that data is protected and to ensure that customer letters being tested are not wrongly sent to real customers. Data could be taken from the mainframe system, but it is also very important to retain integrity of data between different systems. Data that is held in other databases would need to remain consistent with records on the mainframe.

The knowledge of the database structure and which fields need to be depersonalised is likely to lie with the development team – so it is important to consider whether to buy a tool or build it within the organisation.

TDD, ATDD and BDD tools As technology has evolved and processes have improved, new test tools have been developed to support test-driven development (TDD) and behaviour-driven development (BDD) approaches (outlined in **Chapters 2** and **4**). You should be aware that tools exist to support these processes (although many readers may not have used these processes or such tools because they are configured and used primarily by developers during component testing).

TDD, acceptance test-driven development (ATDD) and BDD tools can be further integrated with other types of test tools to design, implement and execute test cases. This integration enables code changes to be compiled quickly, accurately and regularly, and the test suite executed automatically.

TDD tools have been built to support the TDD process, and these tools enable developers to design test cases and test procedures at the component level. These tools usually integrate with or provide interfaces to types of test execution tools.

ATDD is an extension of TDD and, consequently, ATDD tools are an extended version of TDD tools.

BDD tools are also an extension of TDD tools and provide an interface to allow acceptance testers to define user stories or test cases in their own business language (Domain

Specific Language – DSL). The BDD is then configured by the developer to interpret the user story and produce automatically executable test procedures (scripts).

Developers can configure ATDD and BDD tools to enable users to define test cases in a structured format/template that enables such test cases to be captured by ATDD and BDD tools.

USE IN HOTEL CHAIN SCENARIO

As shown in **Figure 6.4**, a TDD development approach (and a TDD tool) can be used in the Agile workstream for the mobile app. Developers can use the CI tool (and configuration management tool) in conjunction with the TDD tool to design test cases and execute frequent tests of daily builds.

Figure 6.4 Testing of daily builds using a set of test tools

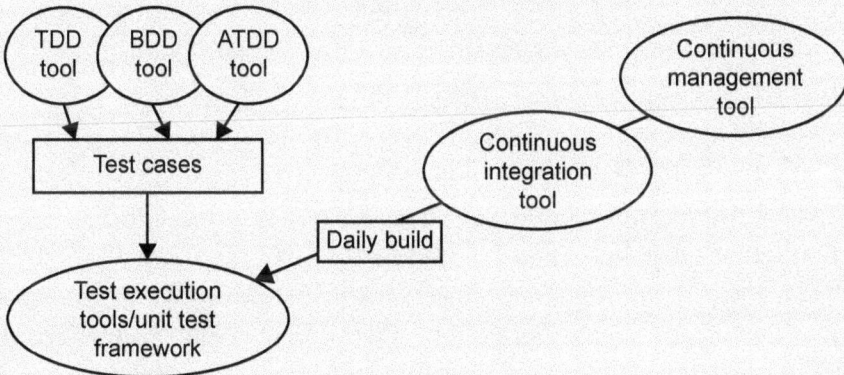

ATDD and BDD templates can be used by acceptance testers to define test cases in a structured version of their own hotel business language. These structured test cases can then be automated and run by test execution tools (or using a unit test framework), which enables the automated part of acceptance testing to be carried out more efficiently.

CHECK OF UNDERSTANDING

1. What types of inputs can a test design tool use to generate test cases?
2. Name four types of management tools that support the test process.
3. In what part of the SDLC are TDD tools most likely to be used?

Many of the tools discussed in the section above ('Test design and implementation tools') are also used for test execution (or can interface with test execution tools).

Test execution and coverage tools

Test execution tools cover a wide range from basic test comparators to ATDD tools (see above) that convert acceptance test cases into executable scripts and then report upon whether they have passed or failed.

Test execution tools Test execution tools allow test scripts to be run automatically (or at least semi-automatically). A test script (written in a programming language or scripting language) is used to navigate through the system under test and to compare predefined expected outcomes with actual outcomes. The results of the test run are written to a test log. Test scripts can then be amended and reused to run other or additional scenarios through the same system. Some tools offer GUI-based utilities that enable amendments to be made to scripts more easily than by changing code. These utilities may include:

- configuring the script to identify particular GUI objects;
- customising the script to allow it to take specified actions when encountering particular GUI objects or messages;
- parameterising the script to read data from various sources.

Record (or capture playback) tools Record (or capture playback) tools can be used to record a test script and then play it back exactly as it was executed. However, a test script usually fails when played back owing to unexpected results or unrecognised objects. This may sound surprising, but consider entering a new customer record onto a system:

- When the script was recorded, the customer record did not exist. When the script is played back the system correctly recognises that this customer record already exists and produces a different response, thus causing the test script to fail.
- When a test script is played back and actual and expected results are compared, a date or time may be displayed. The comparison facility will spot this difference and report a failure.
- Other problems include the inability of test execution tools to recognise some types of GUI control or object. This might be able to be resolved by coding or reconfiguring the object characteristics (but this can be quite complicated and should be left to experts in the tool).

Also note that expected results are not necessarily captured when recording user actions and therefore may not be compared during playback.

While the recording of tests is of particular help with automating GUI-based testing, it can also be useful during exploratory testing for reproducing a defect or for documenting how to execute a test, as the tools capture the exact screen navigation steps that were performed on the system. In both cases, the script can then be made more robust by a technical expert so that it handles valid system behaviours, depending on the inputs and the state of the system under test.

Data-driven testing   Robust test scripts that deal with various inputs can be converted into data-driven tests. This is where hard-coded inputs in the test script are replaced with variables that point to data in a data table. Data tables are usually spreadsheets with one test case per row, each row containing test inputs and expected results. The test script reads the appropriate data value from the data table and inserts it at the correct point in the script (e.g. the value in the Customer Name column is inserted into the Customer Name field on the input screen).

Keyword-driven testing   A further enhancement to data-driven testing is the use of keyword-driven (or action word) testing. Keywords are included as extra columns in the data table. The script reads the keyword and takes the appropriate actions and subsequent path through the system under test. Conditional programming constructs such as IF ELSE statements or SELECT CASE statements are required in the test script for keyword-driven testing.

Programming skills and programming standards are required to use the tool effectively. It may be that these can be provided by a small team of technical experts within the test organisation or from an external company. In data-driven, and particularly keyword-driven, approaches, the bulk of the work can be done by manual testers, with no knowledge of the scripting language, defining their test cases and test data and then running their tests and raising defects as required. However, this relies on robust and well-written test scripts that are easy to maintain. This takes much time and effort before any sort of payback is achieved.

It is essential that time (and consequently budget) is allowed for test scripts to be maintained. Any change to a system can mean that the test scripts need to be updated. Therefore, the introduction of a new type of object or control could result in a mismatch being found between the previous object type and the new one. The relevant level of technical skills and knowledge is required to do this.

USE IN HOTEL CHAIN SCENARIO

Let us assume that a test execution tool is already used for regression testing. Existing automated test scripts could be analysed to identify which ones can be reused and to identify gaps in the coverage for the new enhancement. These gaps could be filled by running cases manually or by writing new automated test scripts. Rather than starting from scratch, it may be possible to produce additional automated scripts by reusing some code or modules already used by existing scripts, or by using parameterisation and customisation utilities. In this enhancement, the automated scripts used to test the unchanged documents could be run without having to be amended.

The automated scripts to produce the amended documents would need to be analysed and updated as required. The navigation part of the script would be largely unchanged, but the comparison between actual and expected results would probably be performed manually the first time round. Once the test has passed manually, the comparison could be added to the script for reuse in the future.

> Automated scripts for new documents could be added to the regression pack after this release is complete.

Figure 6.5 shows how the benefits of using test execution tools take some time to pay back. Note how in the early stages the cost of using automated regression testing is greater than the cost of manual regression testing. This is due to the initial investment, implementation, training, the creation of automated scripts and so on. However, the cost each additional time the test is run is less for automated regression testing than it is for manual regression testing. Therefore, the lines on the graph converge and at a point in time (known as the break-even point) the lines cross. After this point the total cost to date for automated testing is less than the total cost to date for manual regression testing.

Figure 6.5 Test execution tools payback model

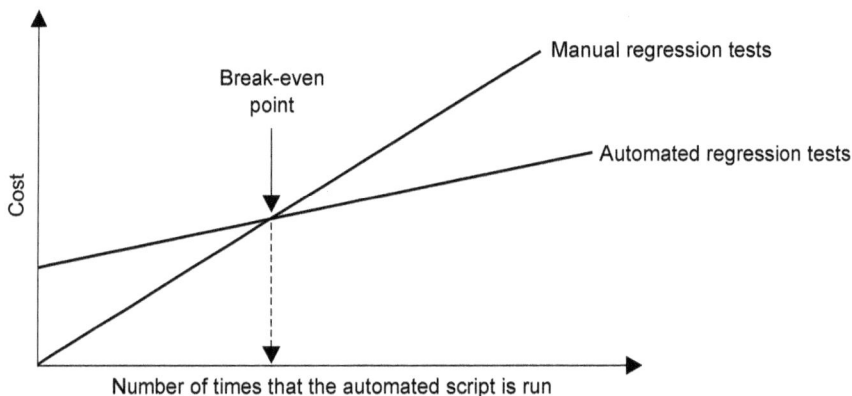

This is clearly a simplistic view, but it demonstrates how an initial investment in test execution tools can be of financial benefit in the medium to long term. There are other, less tangible, benefits as well. However, to get this financial benefit you will need to be sure that there is a requirement to run the same (or very similar) regression tests on many occasions.

Test harnesses and unit test frameworks A test harness is a test environment comprising stubs and drivers needed to execute a test. It is used primarily by developers to simulate a small section of the environment in which the software will operate. Test harnesses are usually written in-house by developers to perform component or integration testing for a specific purpose. Test harnesses often use 'mock objects' known as 'stubs' (which stub out the need to have other components or systems by returning predefined values) and 'drivers' (which replace the calling component or system and drive transactions, messages and commands through the software under test).

Test harnesses can be used to test various systems or objects ranging from a middleware system (as in **Figure 6.6**) to a single or small group of components. They are frequently used in Agile development so that existing tests can be rerun as regression tests to establish whether existing functionality is adversely impacted by the changes made.

A unit test framework is generally more robust and reusable than a standalone test harness and is typically able to support multiple test harnesses for related purposes. It may also provide additional support for the developer, such as debugging capabilities.

USE IN HOTEL CHAIN SCENARIO

Bookings are entered via the web or GUI front-ends and are loaded onto the mainframe. An overnight batch runs on the mainframe and generates XML messages that are then processed by the middleware system, which makes a further call to the mainframe to read other data. The middleware system then generates further XML messages, which are processed by other systems, resulting in the production of letters to customers.

There are several benefits that can be obtained from using a test harness that generates the XML messages produced by the mainframe:

- It would take a lot of time and effort to design and execute test cases on the mainframe system and run the batch.

- It would be costly to build a full environment.

- The mainframe code needed to generate the XML messages may not yet be available.

- A smaller environment is easier to control and manage. It enables developers (or testers) to perform component and integration testing more quickly because it is easier to localise defects. This allows a quicker turnaround time for fixing defects.

The diagram in **Figure 6.6** shows that a test harness has been built using a spreadsheet and macros (the driver) to generate XML messages and send them to the middleware. A stub is used to simulate the calls made by the middleware to the mainframe. The contents of the XML messages produced by the middleware can then be compared with the expected results. This could be enhanced into a more robust and reusable unit test framework that can support additional test harnesses for multiple XML messages.

Figure 6.6 Test harness for middleware

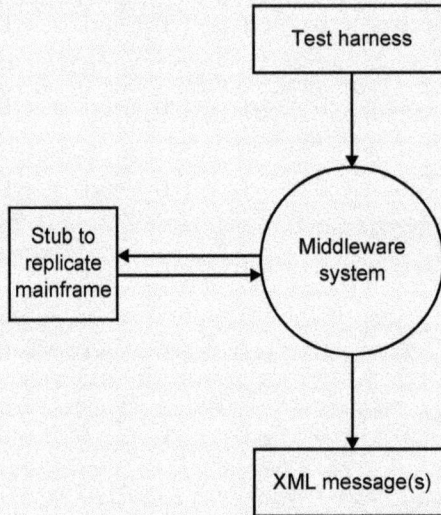

There are similarities with the principle behind data-driven testing using test execution tools because the harness allows many different test cases to be designed and run without the time-consuming process of keying them manually. This raises the question of how much benefit can be obtained from using a test execution tool when a test harness can be used instead. As usual, it depends on the circumstances, the risk, the purpose and the level of testing being performed.

Coverage tools Coverage tools measure the percentage of the code structure covered across white box measurement techniques such as statement coverage and branch or decision coverage. In addition, they can also be used to measure coverage of modules and function calls. Coverage tools are often integrated with static and dynamic analysis tools and are primarily used by developers.

Coverage tools can measure code written in several programming languages, but not all tools can measure code written in all languages. They are useful for reporting coverage measurement and can therefore be used to assess test completion criteria and exit criteria.

Coverage measurement of requirements and test cases/scripts run can usually be obtained from test management tools. This function is unlikely to be provided by coverage tools.

Coverage measurement can be carried out using intrusive or non-intrusive methods. Non-intrusive methods typically involve reviewing code and running code. Intrusive methods, such as 'instrumenting the code' involve adding extra statements into the code. The code is then executed and the extra statements write back to a log in order to identify which statements and branches have been executed.

Instrumentation code can then be removed before it goes into production.

Intrusive methods can affect the accuracy of a test because, for example, slightly more processing will be required to cope with the additional code. This is known as the probe effect and testers need to be aware of its consequences and try to keep its impact to a minimum.

USE IN HOTEL CHAIN SCENARIO

In the hotel chain scenario, assume that the exit criteria for component testing includes the required code coverage shown in **Table 6.2**.

Table 6.2 Exit criteria for component testing

Function	Module risk level	Branch coverage	Statement coverage
BACS	High	100%	100%
Email message to selected customers	Medium	Not specified	100%
Look-up error message/ screen navigation	Low	50%	75%

Non-functional testing tools

These types of tools allow the tester to perform non-functional testing that is difficult or impossible to perform manually. There are many tools that support non-functional testing, so we will limit the discussion to three non-functional test types: performance, security and usability.

Performance testing tools Performance testing is very difficult to do accurately and in a repeatable way without using test tools. Therefore, performance testing tools have been developed to carry out both load testing and stress testing.

Load testing reports on the performance of a system under test, under various loads and usage patterns. A load generator (which is a type of test driver) can be used to simulate the load and required usage pattern by creating virtual users that execute predefined scripts across one or more test machines. Alternatively, response times or transaction times can be measured under various levels of usage by running automated repetitive test scripts via the user interface of the system under test. In both cases output will be written to a log. Reports or graphs can be generated from the contents of the log to monitor the level of performance.

Performance testing tools can also be used for stress testing. In this case, the load on the system under test is increased gradually (ramped up) in order to identify the usage pattern or load at which the system under test fails. For example, if an air traffic control system supports 200 concurrent aircraft in the defined air space, the entry of the 201st or 205th aircraft should not cause the whole system to fail.

Performance testing tools can be used against whole systems, but they can also be used during system integration testing to test an integrated group of systems, one or more servers, one or more databases or a whole environment.

If the risk analysis finds that the likelihood of performance degradation is low, then it is likely that no performance testing will be carried out. For instance, a small enhancement to an existing mainframe system does not necessarily require any formal performance testing. Normal manual testing may be considered sufficient (during which poor performance might be noticed).

There are similarities between performance testing tools and test execution tools in that they both use test scripts and data-driven testing. They can both be left to run unattended overnight and each needs a heavy upfront investment, which will take some period of time to pay back.

Performance testing tools can find defects such as:

- general performance problems;
- performance bottlenecks;
- memory leakage (e.g. if the system is left running under heavy load for some time);
- record-locking problems;
- concurrency problems;
- excess usage of system resources;
- exhaustion of disk space.

The cost of some performance tools is high, and the implementation and training costs are also high. In addition, finding experts in performance testing is not always easy. Therefore, it is worth considering using specialist consultancies to come in and carry out performance testing using such tools.

USE IN HOTEL CHAIN SCENARIO

The likelihood of poor website performance and the cost of lost business and reputation are likely to be sufficient to justify the use of performance testing to mitigate this risk. Performance testing can range from using a relatively cheap tool to indicate whether performance has improved or deteriorated as a result of the enhancement, to an extensive assessment of response times under normal or maximum predicted usage and identification of the usage pattern that will cause the system to fail.

It is likely that performance test tools will have been used when the website was first developed. Therefore, it may be easy to reuse existing tools to do a regression test of performance. If performance tools were not used when the website was developed, it is unlikely to be worthwhile buying and implementing expensive performance testing tools.

An alternative option is to use tools that already exist on servers or in the test environment to monitor performance. It may also be worth considering using external consultants.

Security testing tools   Security testing tools are used to detect security threats and to evaluate the security characteristics of software. A security testing tool is typically used to assess the ability of the software under test to:

- handle computer viruses;
- protect data confidentiality and data integrity;
- prevent unauthorised access;
- carry out authentication checks of users;
- remain available under a denial of service (DOS) attack;
- check non-repudiation attributes of digital signatures.

Security testing tools are mainly used to test ecommerce, ebusiness and websites. For example, a third-party security application such as a firewall may be integrated into a web-based application.

USE IN HOTEL CHAIN SCENARIO

Security testing tools could be used to test that the firewall and other security applications built into the hotel chain's systems can:

- resist a DOS attack;
- prevent unauthorised access to data held within the database;
- prevent unauthorised access to encrypted XML messages containing bank account details.

This work could be carried out by a third-party consultancy that specialises in penetration testing and related services.

The skills required to develop and use security testing tools are very specialised and such tools are generally developed and used on a particular technology platform for a particular purpose. Therefore, it may be worth considering the use of specialist consultancies to perform such testing. For example, specialist consultancies are often engaged to carry out penetration testing. This type of testing is to establish whether malicious attackers can penetrate the organisation's firewall and hack into its systems.

Security testing tools need to be constantly updated because there are problems solved and new vulnerabilities discovered all the time – consider the number of Windows security releases to see the scale of security problems.

Usability testing tools (including accessibility and localisation test tools)   Usability test tools typically record the mouse clicks made by remote usability testers when carrying out a specified task. Some tools also enable other data to be captured, such as screenshots, written notes and voice recordings of verbal comments. This recorded data is generally stored in a database so that it can be analysed easily by staff at the organisation commissioning the testing.

Note that the usability testing tool market is changing very quickly, and new types of usability tools may appear over the next few years. Recent developments include:

- Accessibility test tools – these are an extension of usability test tools. Accessibility testing is defined as testing to determine the ease by which users with disabilities can use a component or system.
- Localisation test tools – these have been developed to support the testing of local-language versions of widely available global software products.

USE IN HOTEL CHAIN SCENARIO

The changes to the website to improve usability could be tested by a specialist usability testing company who employ, say 50, remote users. The remote users would be given a high-level requirement that would exercise the website changes, such as:

- Go to a specified test URL and book three rooms from 3 August to 5 August and two rooms from 7 August. Pay by credit card XYZ.

The mouse clicks, other inputs and comments recorded by the 50 remote users in carrying out this task would be stored in a database and an analysis report produced by the specialist usability testing company for the hotel chain. This analysis could highlight areas of poor usability in the test website, which could be improved before being deployed to the live website.

DevOps tools

As discussed in **Chapter 2**, DevOps promotes team autonomy, fast feedback, integrated toolchains and technical practices such as continuous integration, continuous delivery and continuous deployment. These practices are achieved through extensive use of tools including, from a testing perspective, test automation. Such tools need to be introduced and maintained, and test automation requires additional skilled resources to achieve these goals. There are open source tools and testing frameworks available for DevOps that help organisations to save time with the aim of automating the entire test process.

Collaboration tools
There are many tools that support communication and information sharing, especially in Agile.

Wikis allow teams to build and share an online knowledge base covering topics such as:

- product feature diagrams, prototype diagrams and whiteboard discussion photos;
- metrics and charts on dashboard status;
- useful techniques, tools or ideas;
- capture of electronic messages or conversations between members deemed useful to others.

Desktop sharing tools are of particular benefit in a distributed environment and can be used in reviews and to capture product demonstrations, which can also be posted to the team's wiki.

Instant messaging, audio teleconferencing and video chat tools provide real-time communication between team members, especially in a distributed environment.

Virtual whiteboarding tools can be used for brainstorming, retrospectives, debugging, design work, etc.

Tools support for scalability and deployment standardisation
One example under this category is virtual machine (virtualisation) tools, which emulate additional operating systems on a computer to run programs and deploy apps in other windows. They allow a single physical resource (e.g. a server) to operate as many separate, smaller resources. These tools can, for example, run a Windows instance on macOS or vice versa, as well as different operating system combinations such as ChromeOS or Linux. They can be of great benefit in the higher test levels, such as system testing, as they allow test teams to parallel test the application without impacting others' test data.

Containerisation tools also fall under this category. Containerisation is a process that packages an application's code with all the files and libraries it needs to run on any infrastructure. These tools create a single lightweight executable, which is also referred to as a 'container'. They provide scalability and portability as they deploy applications in multiple environments without rewriting the program code, and allow developers to add multiple containers for different applications on a single machine (using virtualisation).

Other tools that are not designed specifically for testers or developers can also be used to support one or more test activities. These include:

- spreadsheets;
- word processors;

- email;
- back-up and restore utilities;
- SQL and other database query tools;
- project planning tools;
- debugging tools (although these are more likely to be used by developers than testers).

For example, in the absence of test management or defect management tools, defects could be recorded on word processors and could be tracked and maintained on spreadsheets. Tests passed or failed could also be recorded on spreadsheets.

USE IN HOTEL CHAIN SCENARIO

Other software tools could also be used:

- A spreadsheet could be used for producing decision tables or working out all the different test scenarios required. It could also be used to manipulate test management information so that it can be presented in the format required in weekly or daily test progress reports.
- Word processors could be used for writing test strategies, test plans, weekly reports and other test deliverables.
- Email could be used for communicating with developers about defects and for distributing test reports and other deliverables.
- Back-up and restore utilities could be used to restore a consistent set of data into the test environment for regression testing.
- SQL could be used for analysing the data held in databases in order to obtain actual or expected results.
- Project planning tools could be used to estimate resources and timescales, and to monitor progress.
- Debugging tools can be used by developers to localise and fix defects.

Benefits and risks of test automation

While tools can provide significant benefits for testing, they do not guarantee success. They must be carefully selected and introduced, and require ongoing support and maintenance to achieve cost-effectiveness. It is important to consider the potential benefits and risks of introducing a tool into a project or organisation. While the syllabus suggests this refers to test automation, it applies to all types of tools that support the test process.

Potential benefits

- Time saved by reducing repetitive manual work. The main benefit of using test tools is similar to the main benefit of automating any process. That is, the amount of time and effort spent performing routine, mundane, repetitive tasks is greatly reduced. Regression testing is a prime candidate for automation, especially in Agile and other iterative life cycle methodologies where time constraints necessitate, for example, automated rather than manual testing of previous iteration software. As discussed earlier, the data-driven testing scripting technique is used in automation to automatically enter large and varied volumes of predefined input data at runtime, significantly reducing manual input effort.

- Prevention of simple human errors through greater consistency and repeatability. Automation of any process usually results in more predictable and consistent results. Similarly, the use of test tools provides more predictable and consistent results as human failings – such as manual keying errors, misunderstandings, incorrect assumptions, forgetfulness and so on – are eliminated. It also means that any reports or findings tend to be objective rather than subjective. For instance, humans often assume that something that seems reasonable is correct, when in fact it may not be what the system is supposed to do.

- More objective assessment (e.g. coverage) and providing measures that are too complicated for humans to derive. Examples of this are tools that measure structural coverage during component testing, such as the percentage of branch test coverage achieved by a suite of component tests, providing the developer with an indication of gaps that need addressing through additional test design. Static analysis tools can also measure code complexity to indicate increased risk of failure and potential maintainability issues. One example is cyclomatic complexity, which is a quantitative measure of the number of linearly independent paths through a program's logic, obtained from analysis of the edges and nodes of a control flow graph. To attempt to measure structural coverage or code complexity manually would be both very time-consuming and error-prone.

- Easier access to information about testing to support test management and test reporting. The widespread use of databases to hold the data input, processed or captured by the test tool, means that it is generally much easier and quicker to obtain and present accurate test management information, such as test progress, defects found/fixed and so on (see **Chapter 5**). The introduction of web-based tools that have databases stored in the cloud means that such information is available to global organisations 24 hours per day, 7 days per week. This facilitates round-the-clock working and can reduce elapsed times to analyse, fix and retest defects.

- Reduced test execution times to provide earlier defect detection, faster feedback and faster time to market. An example of this is CI, used extensively in Agile and DevOps, whereby developers run automated tests at the component and component integration levels as part of the, often daily, CI process This achieves the benefits of detecting defects much earlier in the testing life cycle and sends quick feedback to the team on the quality of the code. In turn, higher-quality code is produced early, facilitating the potential to deliver to the market on a much faster basis than seen in traditional sequential life cycles.

- More time for testers to design new, deeper and more effective tests. This time saved through automation can not only reduce the cost of test execution, but also allow testers to spend more time on the more intellectual tasks of test planning, analysis and design. In turn, this can enable more focused and appropriate testing to be done rather than having many testers working long hours and running hundreds of tests.

Potential risks

- Unrealistic expectations about the benefits of a tool (including functionality and ease of use). Most of the risks associated with the use of test tools are concerned with over-optimistic expectations of what the tool can do and a lack of appreciation of the effort required to implement and obtain the benefits that the tool can bring. For example, consider the production environments of most organisations thinking about using test tools. They are unlikely to have been designed and built with test tools in mind. Therefore, assuming that you want a test environment to be a copy of production (or at least as close to it as possible), you will also have a test environment that is not designed and built with test tools in mind. Consider the test environments used by vendors to demonstrate their test tools. If you were the vendor, would you design the environment to enable you to demonstrate the tool at its best or to demonstrate the shortcomings it may encounter in a typical test environment? Therefore, unless detailed analysis and evaluation is done, it is likely that test tools will end up as something that seemed a good idea at the time but have been largely a waste of time and money.

- Inaccurate estimations of time, costs, effort required to introduce a tool, maintain test scripts and change the existing manual test process. After a test tool has been implemented and measurable benefits are being achieved, it is important to put in sufficient effort to maintain the tool, the processes surrounding it and the test environment in which it is used. Otherwise there is a risk that the benefits being obtained will decrease and the tool will become redundant. Additionally, opportunities for improving the way in which the tool is used could be missed. For example, the acquisition of various test tools from multiple vendors will require interfaces to be built or configured to import and export data between tools. Otherwise much time may be spent manually cutting and pasting data from one tool to another. If this is not done, then data inconsistencies and version control problems are likely to arise. Similar problems may arise when testing with third-party suppliers or as a result of mergers and acquisitions. The increase in common standards for interfaces such as XML means that the capability for developing successful interfaces is greater, but substantial time and effort are often still required. Maintenance effort will also be required to upgrade and reconfigure tools so that they remain compliant with new platforms or operating systems.

- Using a test tool when manual testing is more appropriate and relying on a tool too much (e.g. ignoring the need for human critical thinking). While tools can have a great appeal (for many, but not all) it can be tempting to overrely on them, especially in the field of test automation. Not all testing can, or should, be automated. Some test techniques such as exploratory testing, test types such as usability testing and test levels such as acceptance testing are predominantly manual activities as they require human inspection of the test object while it is executing. Furthermore, runtime failures are often observed by chance. For example, while manually executing a test case that is designed to test the error handling of an invalid user password at login, it is discovered (by the tester) that the user is unable to tab between the user ID and password fields, an accessibility failure that was not the intent behind the design of the test but nonetheless is significant for those that rely upon tabbing, such as those who are visually impaired.

- The dependency on the tool vendor, which may go out of business, retire the tool, sell the tool to a different vendor or provide poor support (e.g. responses to queries, upgrades and defect fixes). This is one of the more difficult risks to mitigate, as we cannot directly control the actions of a tool vendor. However, one of the important steps in introducing a tool into a project or organisation is to research the tool vendor before purchase. How financially stable are they? What is their roadmap for the tool – for example, would you buy it if they plan to replace it with a newer product in six months' time? The same principle applies to appliances such as smartphones. Would you enter a lengthy fixed contract on a model of phone, knowing that a new model is due on the market imminently and offers better features? Make sure you understand the support and maintenance terms offered by the vendor and what is included in the purchase price. See what other users of the vendor say about the product as well as their experience of the vendor's support.

- Using open source software, which may be abandoned, meaning that no further updates are available, or its internal components may require quite frequent updates as further development takes place. Open source tools have become extremely popular these days; indeed, some products are seen to be superior to more expensive commercial offerings. However, support for such tools tends to be informal, user-forum supplied. There is no contractual obligation for the developers of the tool to continue to maintain it or provide support for it. Many open source tools are quite technical in nature, which can increase the support required. Many have no training available or training supplied by experienced users, and it may be difficult to find users or consultants who know how to use the tool. In short, the project carries the risk.

- The automation tool is not compatible with the development platform. Automation tools, in particular, can be very 'platform' specific. Spending a lot of money on a commercial test automation tool, for example, only to discover that it doesn't work on the current development platform would not please financial stakeholders. This is why a proof-of-concept is an essential step in the tool's introduction phase. In essence, 'try before you buy' and make sure it functions as required on the current development platform. Furthermore, will it continue to function on future development platforms that your organisation is planning to, or may, be involved in? Automation tools, like many tools, tend to be organisational assets as they are expected to have a long 'shelf life'. Indeed, some automation tools may not pay back the initial purchase cost during the lifetime of the current project, given the costs of getting it up and running for the first time. Therefore, it is essential to look beyond its current intended deployment.

- Choosing an unsuitable tool that did not comply with the regulatory requirements or safety standards. This is a particular risk when selecting open source tools, where according to many safety-critical standards, such as (RTCA) DO-178C 'Software considerations in airborne systems and equipment certification', tools used in development according to the standard must be certified. Commercially supplied tools tend to be certified by the vendor, whereas certifying open source tools may be the responsibility of the organisation using them.

After a test tool has been implemented and measurable benefits are being achieved, it is important to put in sufficient effort to maintain the tool, the processes surrounding it and the test environment in which it is used. Otherwise there is a risk that the benefits being obtained will decrease and the tool will become redundant. Additionally, opportunities for improving the way in which the tool is used could be missed. This is particularly important in a DevOps environment as continuous improvement is very important.

SUMMARY

We have seen that the main benefits of using test tools are generally the same as the benefits of automating a process in any industry: time saved and predictable and consistent results.

However, we have also seen that there can be considerable costs in terms of both time and money associated with obtaining such benefits. The point at which the use of tools becomes economically viable depends on the amount of reuse, which is often difficult to predict.

Other risks include over-optimistic expectations of:

- what the tool can do;
- how easy it is to use;
- the amount of maintenance required.

We have seen that there are many types of test tools and that they provide support to a variety of activities within the test process. We have also seen that tools are used by a variety of staff in the software development process and that some are of greater benefit to developers than testers.

Example examination questions with answers

E1. K1 question
A project requires test tools to support test case creation and to measure code complexity. Which two of the following classes of test tool are appropriate to select tools from?

    i.   Management tools.
    ii.   Test design and implementation tools.
    iii. Test execution and coverage tools.
    iv. Static testing tools.
    v.   Collaboration tools.

    A. ii and iv.
    B. ii and iii.
    C. iii and iv.
    D. i and v.

E2. K2 question
Which of the following correctly identifies a benefit of test automation?

    A. Version control of test assets is no longer required.
    B. Greater consistency and repeatability of tests.
    C. The tool vendor is always available for help and advice.
    D. Regression testing will not be needed.

E3. K1 question
Which of the following is a risk of using test tools?

    A. Unrealistic expectations about the benefits tools can bring.
    B. Tools are less objective in assessing coverage than humans.
    C. The time spent on test automation can result in late shipment to the market.
    D. Automation tools may not be compatible with project management tools.

Answers to the self-assessment questions in the chapter

SA1. The correct answer is D.

SA2. The correct answer is C.

SA3. The correct answer is A.

Answers to example examination questions

E1. The correct answer is A.

The correct classes of tool are test design and implementation tools (test design tool for test cases) and static testing tools (to enable the measurement of code complexity). The only answer that identifies this combination is option A.

E2. The correct answer is B. Option B is specifically listed as a potential benefit of test automation and is the correct answer.

    A.   is incorrect; neglect of version control is a potential risk.

    C.   is incorrect; vendors going out of business or discontinuing support of a tool are potential risks.

    D.   is incorrect; regression testing may be easier and quicker to do but will still be required.

E3. The correct answer is A. Most of the risks associated with the use of test tools are concerned with over-optimistic expectations of what the tool can do.

    B.   is incorrect as tools are **more** objective in their assessment of coverage.

    C.   is incorrect as reduced test execution times provide earlier defect detection, faster feedback and faster time to market.

    D.   is incorrect as there is no need for test automation tools to be compatible with project management tools. They do, however, need to be compatible with the development platform.

7 THE EXAMINATION

Marie Salmon

THE EXAMINATION

The examination structure

The Certified Tester Foundation Level (CTFL) examination is a one-hour examination made up of 40 multiple-choice questions. These are the main aspects to the examination's structure:

- The questions are all equally weighted and earn one point each.
- Questions are set from learning objectives stated in each section.
- The number of questions associated with each section of the syllabus is in proportion to the amount of time allocated to that section of the syllabus, which roughly translates into:
  - Section 1: 8 questions;
  - Section 2: 6 questions;
  - Section 3: 4 questions;
  - Section 4: 11 questions;
  - Section 5: 9 questions;
  - Section 6: 2 questions.
- Each question will examine the cognitive level of knowledge specified for its learning objective, known as 'K levels'.
  - **Level 1: Remember (K1)** – the candidate will remember, recognise and recall a term or concept.
  - **Level 2: Understand (K2)** – the candidate can select the reasons or explanations for statements related to the topic, and can summarise, compare, classify and give examples for the testing concept.
  - **Level 3: Apply (K3)** – the candidate can carry out a procedure when confronted with a familiar task, or select the correct procedure and apply it to a given context.

- The number of questions for each cognitive level of knowledge will be as follows:
  - K1: 20 per cent, 8 questions;
  - K2: 60 per cent, 24 questions;
  - K3: 20 per cent, 8 questions.
- The pass mark is 26 correct answers and there are no penalties for incorrect answers.

The question types

All questions will contain a 'stem', which states the question and four optional answers. Only one of the optional answers will be correct. The remainder can be expected to be plausibly incorrect, which means that anyone knowing the correct answer will be unlikely to be drawn to any of the incorrect answers, but anyone unsure of the correct answer will be likely to find one or more alternatives equally plausible.

Questions will be stated as clearly as possible, even emphasising keywords by emboldening where this will add clarity. Questions will be set to test your knowledge of the content of the topics covered in the syllabus and not your knowledge of the syllabus itself.

K1 questions require the candidate to **remember** a topic and will be written in a style similar to the question shown in the next box.

EXAMPLE OF A K1 QUESTION

(This one is taken from **Chapter 3**.)

In reviews, what are the main responsibilities of the 'Author' role?

        A.   Record anomalies from reviewers and any decisions made.
        B.   Provide resources, such as the staff and time for the review.
        C.   Ensure effective running of the review meeting.
        D.   Create and fix the work product under review.

(The correct answer is D.)

K2 questions may be of the same type as the K1 example, but with a more searching stem to demonstrate **understanding**. Another type of question is known as the 'Roman type'. This is particularly well suited to questions involving comparisons or testing the candidate's ability to identify correct combinations of information. The example in the next box is a K2 question of the Roman type.

EXAMPLE OF A K2 QUESTION

(This one is taken from **Chapter 4**.)

Which of the following statements are correct for exploratory testing?

  i.   It is used to learn more about the test object.
  ii. It uses a fault attack approach.
  iii. All participants have defined roles.
  iv. It uses a test charter to guide the testing.
  v.   It is useful when there is time pressure on testing.

  A.   i, ii and v are correct.
  B.   ii and iii are correct.
  C.   i, iv and v are correct.
  D.   i, iii and iv are correct.

(The correct answer is C.)

K3 questions test the candidate's ability to **apply** a topic. The next box gives an example of a K3 question using the 'Matching type' of question. This format provides a way to connect a word, sentence or phrase in one list to a corresponding word, sentence or phrase in the second list.

EXAMPLE OF A K3 QUESTION

(This one is taken from **Chapter 5**.)

You are to execute a regression test suite, consisting of 100 test cases, for the fifth release of the current project. The test suite has had minimal change over the previous releases and the effort estimated and captured for each release was as follows:

Release	Estimate	Actual
1	70 hours	80 hours
2	80 hours	50 hours
3	65 hours	50 hours
4	60 hours	Not captured

The actual effort for Release 4 for was not captured correctly, so only the estimate can be used.

Applying the extrapolation estimation technique against the previous four releases, what would be the likely effort required for Release 5?

A. 68.75 hours

B. 60 hours

C. 45 hours

D. 50 hours

The correct answer is B, 60 hours. The extrapolation technique works by averaging the actual effort for the previous releases/iterations. For Release 4 we can only use the estimate, so it will be 80 + 50 + 50 + 60 = 240/4 = 60 hours. You can also calculate that this technique has been used for the previous four iterations.

Option A averages all four estimates, not the actuals.

Option C disregards the Release 4 estimate (so only averages the first three releases – that is, 180/4).

Option D just takes the last two actuals as the estimate.

Remember that K1, K2 and K3 does not equate to easy, moderate or hard. The K level identifies the cognitive level of knowledge being tested, not the difficulty of the question. It is perfectly possible to find K2 questions that are more difficult (in the sense of being more challenging to answer) than a K3 question. Please also note that every question has the same value (1 point) regardless of the K level being tested; any 26 correct answers will guarantee a pass.

Sample examination

Sample examination papers are available from the ISTQB website. They are designed to provide guidance on the structure of the paper and the `rubric' (the rules printed on the front of the paper) of the real examination. The questions in the sample papers will include examples of the various types of questions, so that candidates are aware of the kinds of questions that can arise. Any topic or type of question in the sample papers can be expected to arise in a real examination. Bear in mind that the sample papers may change from time to time to reflect any changes to the syllabus or in the way questions are set.

Examination technique

In a relatively short examination there is little time to devote to studying the paper in depth. However, it is wise to pause before beginning to answer questions, while you assimilate the contents of the question paper. This brief time of inactivity is also a good opportunity to consciously slow down your heart rate and regulate your breathing;

nervousness is natural, but it can harm your performance by making you rush. A few minutes spent consciously calming down will typically be advantageous. There will still be enough time to answer the questions; a strong candidate can answer 40 questions in less than 45 minutes.

When you do start, go through the whole paper answering those questions that are straightforward and for which you know the answer. When you have done this, you will have a smaller task to complete and you will probably have taken less than a minute for each question that you have already answered, giving you more time to concentrate on those that you will need more time to answer.

Next, turn to those you feel you understand but that will take you a little time to work out the correct answer, and complete as many of those as you can. The questions you are left with now should be those you are uncertain about, and you can take a little more time over each of them.

REVISION TECHNIQUES

There are some golden rules for exam revision:

- Do as many example questions as you can so that you become familiar with the types of questions, the way questions are worded and the levels (K1, K2, K3) of questions that are set in the examination.

- Be active in your reading. This usually means taking notes, but this book has been structured to include regular checks of understanding that provide you with prompts to ensure that you have remembered the key ideas from the section you have just revised. In many cases, information you need to remember is already in note form for easy learning.

- One important way to engage with the book is to work through all the examples and exercises. If you convince yourself you can do an exercise, but you do not actually attempt it, you will only discover the weakness in that approach when you are sitting the exam.

- Learning and revision need to be reinforced. There are two related ways to do this:

  - Make structured notes to connect related ideas. This can be done via lists, but a particularly effective way to make the connections is by using a technique known as mind mapping (there are many free tools on the internet or, if you find you really like the technique, you can invest in a more feature-rich product).

  - Return to a topic that you have revised to check that you have retained the information. This is best done the day after you first revised the topic and again a week after, if possible. If you begin each revision section by returning to the 'Check of understanding' boxes in some or all of the chapters you worked with in previous revision sessions, it will help to ensure you retain what you are revising.

- Read the syllabus and become familiar with it. Questions are raised directly from the syllabus and often contain wording similar to that used in the syllabus. Familiarity with the syllabus document will more than repay the time you will spend gaining that familiarity.

REVIEW

The layout, structure and style of this book are designed to maximise your learning: by presenting information in a form that is easy to assimilate; by listing things you need to remember; by highlighting key ideas; by providing worked examples; and by giving exercises with solutions. All you need for an intense and effective revision session is in these pages.

The best preparation for any examination is to practice answering as many realistic questions as possible. This is one way to use the ISTQB sample paper and questions included at the end of each chapter in this book. Once you feel your revision is complete, a good check of your readiness for tackling the real examination is to take a 'mock exam' under conditions as close to the real examination as possible. Set a timer and work through the questions using the same approach you plan to use for the actual exam. In the appendices of this book, there is a mock exam, including the answers so that you can see how well you did and a full commentary so that you can identify where you went wrong in any questions and revise the topic further.

Good luck with your Foundation Certificate examination.

A1 MOCK CTFL EXAMINATION

Question 1
Which of the following statements correctly describes the relation between testing and quality assurance?

    A. Testing is a preventive approach that focuses on improvement of processes, quality assurance is a form of quality control.

    B. Testing is a form of quality control, quality assurance applies to both development and testing processes.

    C. Testing will find defects in a product, quality assurance is used to fix defects found in testing.

    D. Testing and quality assurance are different terms used to describe the same activity.

Question 2
What is an advantage of the whole team approach?

    A. It enables a project to use members of the public for testing.

    B. It provides a high level of test independence.

    C. It allows various skills sets within a team to be leveraged.

    D. It makes the developer responsible for quality.

Question 3
How does the test management role differ from the testing role?

    A. It always has the same activities and tasks assigned, regardless of context.

    B. It takes overall responsibility for the engineering (technical) aspect of testing.

    C. It is not required for projects following Agile software development methods.

    D. It is mainly focused on test planning, test monitoring and control and test completion.

Question 4
Which of the following correctly identifies a key role and associated responsibilities for a review?

    A. The manager assigns staff and executes control decisions in the event of inadequate outcomes.

    B. The review leader assigns staff and takes overall responsibility for the review.

    C. The reviewers, who must be subject-matter experts, identify potential defects.

    D. The moderator decides who will be involved and organises where and when a meeting will take place.

Question 5
Of the following, which is the most suitable test basis for component testing?

    A. A code specification.

    B. A database module.

    C. An interface definition.

    D. A business process.

Question 6
Which of the following statements *best* describes the role of testing?

    A. Testing ensures that the right version of code is delivered.

    B. Testing can be used to assess quality.

    C. Testing improves quality.

    D. Testing shows that the software is error-free.

Question 7
Which of the following *best* demonstrates the value of static testing?

    A. Project A was late starting testing and subsequently overran the project target and budget.

    B. Project B was late starting development because detailed requirements reviews were held but later completed on time and on budget.

    C. Project C involved all testers in reviewing test specifications and subsequently completed late and over budget.

    D. Project D overran because a defect occurred late in development, resulting in long delays.

Question 8
Which of the following is *not* a test-first approach to development?

    A.   Decision table-driven development (DTDD).

    B.   Acceptance test-driven development (ATDD).

    C.   Test-driven development (TDD).

    D.   Behaviour-driven development (BDD).

Question 9
Which of the following is a common test metric?

    A.   Number of testers in a test team.

    B.   Number of test cases prepared.

    C.   Number of test cases run/not run.

    D.   Number of requirements to be tested.

Question 10
Which of the following are impacts that a software development life cycle (SDLC) could have on testing?

    i.   Cost of delivering the project.

    ii.   Level of detail of test documentation.

    iii.   Choice of test techniques and test approach.

    iv.   Extent of test automation.

    v.   Number of testers required.

    A.   ii, iii and iv.

    B.   i, ii and iv.

    C.   iii, iv and v.

    D.   i, iii and v.

Question 11
Which of the following is *not* a good testing practice?

    A.   Projects prioritise the funding of testing and resourcing to ensure quality.

    B.   Different test levels have specific and different test objectives.

    C.   Testers are involved in reviewing work products as soon as they are available.

    D.   For every software development activity there is a corresponding test activity.

Question 12
Which of the following correctly explains an aspect of the shift-left approach?

    A. It relies solely on static testing and static analysis.

    B. Non-functional testing should be performed as late as possible.

    C. It results in a higher cost of testing but also higher quality.

    D. It does not wait for code to be implemented or components integrated.

Question 13
Which of the following five test techniques match each type of test technique?

    1. Error guessing.

    2. Equivalence partitioning.

    3. Statement testing.

    4. Boundary value analysis.

    5. Branch testing.

    a. Black box test technique.

    b. White box test technique.

    c. Experience-based test technique.

    A. 1c, 2a, 3b, 4a, 5b.

    B. 1a, 2b, 3c, 4b, 5b.

    C. 1c, 2a, 3b, 4b, 5c.

    D. 1b, 2b, 3a, 4c, 5a.

Question 14
Which of the following statements about branch testing is correct?

    A. The aim is to design test cases that exercise statements in the code.

    B. Branch testing is best performed at the end of the SDLC.

    C. Branches in the code can be conditional or unconditional.

    D. It is not possible to achieve 100 per cent branch coverage.

Question 15
When explaining the value of white box testing, which of the following statements is _false_?

    A. It enables stakeholders to see the functionality early in the SDLC.

    B. White box techniques can be used in static testing.

    C. It can be used for code that is not ready for execution.

    D. It provides an objective measurement of code coverage.

Question 16
Which of the following correctly explains error guessing?

    A.   Error guessing is based on testing for the types of errors that developers tend to make.

    B.   Error guessing is a technique for creating lists of mistakes, defects and failures that can be used by developers to avoid mistakes in the future.

    C.   Error guessing is based solely on the tester's experience of past mistakes.

    D.   Error guessing uses evidence from defect reports to ensure that the defect reported is no longer present.

Question 17
Which of the following is a valid entry criterion for a testing phase?

    A.   The project has paid for the full test effort.

    B.   A suitable test environment is available.

    C.   All previous phases of testing have been completed.

    D.   The defects reported in previous phases have been cleared.

Question 18
Which of the following lists of activities are correctly sequenced for a product review process?

    A.   Distributing the work product, estimating effort and timescale, reviewing the work product, creating defect reports, analysing potential defects.

    B.   Estimating effort and timescale, distributing the work product, gathering metrics, analysing potential defects, creating defect reports.

    C.   Selecting reviewers, explaining scope and objectives, noting potential defects, evaluating quality characteristics, fixing defects found.

    D.   Identifying the review type, distributing the work product, analysing potential defects, fixing defects, communicating identified defects.

Question 19
Which of the following acceptance criteria uses the scenario-oriented format?

    A.   Bullet point verification list.

    B.   Given–when–then.

    C.   Input–output mapping.

    D.   As a/I want/so that I can.

Question 20
Which of the following statements about exploratory testing is correct?

A. In exploratory testing, predefined tests are executed, logged and evaluated dynamically during test execution.

B. Exploratory testing is strongly associated with model-based test strategies and can incorporate other black box techniques.

C. Exploratory testing is used to learn more about the test object, to explore it more deeply with focused tests and to create tests for untested areas.

D. Exploratory testing always uses session-based testing to ensure that tests are documented.

Question 21
Which of the following activities are part of test execution?

A. Identifying suitable test techniques, determining testing tasks and drawing up a test schedule.

B. Creating test suites from the test procedures, building the test environment and preparing test data.

C. Comparing actual results with expected results, running tests manually or using tools, and logging the outcome of tests run.

D. Examining the test basis, evaluating the quality of the test basis and identifying and prioritising test conditions.

Question 22
Which of the following test types are performed in which quadrant?

1. Smoke tests and non-functional tests.
2. Functional tests such as user story tests.
3. Component and component integration tests.
4. Exploratory testing and usability testing.

a. Quadrant 1 (technology-facing, support the team).
b. Quadrant 2 (business-facing, support the team).
c. Quadrant 3 (business-facing, critique the product).
d. Quadrant 4 (technology-facing, critique the product).

A. 1b, 2c, 3a, 4d.
B. 1d, 2b, 3a, 4c.
C. 1a, 2d, 3b, 4c.
D. 1a, 2b, 3d, 4c.

Question 23
A project is producing an estimate following the three-point estimation technique.

- The most optimistic estimation (a) is 10 days.
- The most likely estimation (m) is 20 days.
- The most pessimistic estimation (b) is 30 days.

Using the most popular version of the three-point estimation technique, which of the following is the final estimate (E)?

    A.   15 days.
    B.   27 days.
    C.   30 days.
    D.   20 days.

Question 24
Which of the following tools would be used during test execution?

    A.   Static analysis tools.
    B.   Test-driven development tools.
    C.   Collaboration tools.
    D.   Coverage tools.

Question 25
Which of the following is a complete definition of how configuration management supports testing?

    A.   It ensures that testware and system components are uniquely identified.
    B.   It identifies, controls and tracks the system components, the testware and the relationship between them.
    C.   It ensures that all identified documents are referenced unambiguously in test documentation.
    D.   It ensures that all test items are version controlled, tracked for changes and related to each other.

Question 26
A company rewards its sales people on the basis of their sales in each month. Sales staff who make sales worth more than £10,000 are paid the highest bonus, followed by those who earn more than £8,000, those who earn more than £5,000 and those who earn more than £3,000, with each group earning a bonus of 1 per cent less than the group above. Sales staff who earn £3,000 or less are paid no bonus for that month. What is the minimum number of test cases required to cover all valid equivalence partitions for calculating the sales bonus?

    A. 6
    B. 5
    C. 4
    D. 3

Question 27
Which of the following is the best reason to maintain traceability between the test basis and test work products?

    A. Traceability ensures that all requirements have been tested.
    B. Traceability assists in the impact analysis of potential changes.
    C. Traceability enables the project to be delivered on time.
    D. Traceability highlights who is to blame when defects are found.

Question 28
A central heating system timer is calibrated in hours and minutes, using a 24-hour clock. The system allows up to four time zones for each day. On a particular day the system is set to switch on from 06.00 to 08.45, from 11.45 to 13.15, and from 16.45 to 22.45. Using a two-point boundary value system, which of the following times are needed to test this day's functionality?

    A. 06.00, 06.01, 08.45, 08.46, 11.45, 11.46, 13.15, 13.16, 16.45, 16.46, 22.45, 22.46.
    B. 05.59, 06.00, 08.44, 08.45, 11.44, 11.45, 13.14, 13.15, 16.44, 16.45, 22.44, 22.45.
    C. 06.00, 06.01, 08.44, 08.45, 11.44, 11.45, 13.14, 13.15, 16.45, 16.46, 22.44, 22.45.
    D. 05.59, 06.00, 08.45, 08.46, 11.44, 11.45, 13.15, 13.16, 16.44, 16.45, 22.45, 22.46.

Question 29
A requirements document for a new supermarket checkout system has been produced. It is 743 pages long and contains an overall description of the system, detailed workflows and use cases, with mocked up screenshots of specific functions, and an outline of the proposed approach to development. Which of the following is a key success factor for a successful review of this document?

    A.   The review team must include designers, developers and testers, but the document is too technical for users to review.

    B.   The document should be reviewed in small chunks so that reviewers do not lose concentration.

    C.   The review must take the form of an inspection so that detailed metrics are available.

    D.   The review should be arranged at the earliest possible opportunity so that the development team is not held up.

Question 30
Which of the following is a trigger for maintenance testing?

    A.   A new feature is required for an iteration.

    B.   A fix is required before the system goes live.

    C.   The system functionality has been descoped to hit a deadline.

    D.   Data is being migrated for a live system to a different platform.

Question 31
A system is being developed to manage the operation of a warehouse-based robot that collects items from shelves and takes them to a packaging area. The location in which the robot will operate also has human operators working, though in a separate area of the warehouse. How should a product risk analysis inform the testing?

    A.   By helping to determine the particular types and levels of testing to be performed.

    B.   By defining the specific safety features to be incorporated.

    C.   By mandating the use of particular test tools.

    D.   By mandating that all testers are experienced in safety-related systems.

Question 32
Which of the following statements about communicating the status of testing is correct?

    A.   Only one option for communicating should be used per project.

    B.   Informal communication options are best used for distributed teams.

    C.   Communication should be tailored for different stakeholders.

    D.   Verbal communication with team members is not recommended.

Question 33

A suite of tests has been run and some changes have been made to the relevant modules that have affected the priority of the tests. The original priorities were as follows:

Test case	Priority
1	H
2	M
3	H
4	L
5	L
6	H

After the initial tests some remedial work was done, and this work has changed the dependencies between the test cases as follows:

- Test case 3 is now dependent on test case 5.
- Test case 2 is now dependent on test case 3.

What should be the sequence of tests in the test execution schedule?

> A. 1, 3, 6, 2, 4, 5.
> B. 1, 6, 5, 3, 2, 4.
> C. 1, 5, 3, 2, 6, 4.
> D. 1, 6, 3, 2, 5, 4.

Question 34

Which of the statements about the state transition diagram and table of test cases is true?

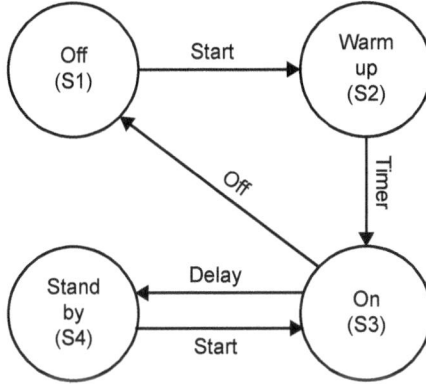

Test case	1	2	3	4	5	6
Start state	S1	S2	S3	S4	S3	S4
Input	Start	Timer	Delay	Start	Off	Off
Expected end state	S1	S2	S3	S4	S3	S4

A. The given test cases cover all valid transitions and at least one invalid transition in the state transition diagram.

B. The given test cases test all valid transitions but not invalid transitions in the state transition diagram.

C. The given test cases test only some of the valid transitions in the state transition diagram.

D. The given test cases represent only some of the valid transitions and all invalid transitions in the state transition diagram.

Question 35

A defect report has been raised for a system that separates apples from a conveyor belt into different sizes and diverts them to appropriate belts. The test found that the system correctly channelled apples below a certain size but failed to detect larger apples and divert them to their own belt.

The dated defect report has a title and provides a short summary of the defect, the degree of risk involved and the severity allocated to the defect. It identifies the test script that was run and the expected result of the test, which was that apples above 10 cm in diameter should be diverted to belt 4.

Which of the following information, if any, is the *most* important item missing from this defect report?

    A. Recommendations.
    B. Actual result.
    C. Priority to fix.
    D. Author.

Question 36

Your project is following a test-first approach and you have been asked to write acceptance tests for the following user story:

User Story: 001

As a bank customer, I want to manage my direct debits using the mobile banking app, so that I can control my monthly bills.

Acceptance criteria:

- I can see a list of all direct debits on my account, select a direct debit and cancel it.

- I can select a direct debit and see the last time it was paid and the amount debited.

Which of the following cases is a positive ATDD test case for this user story?

    A. Login to the mobile banking app, select the direct debit list, select a direct debit, click cancel, cancellation confirmation message is displayed on screen.
    B. As a bank customer, I want to change my home address and phone number, using the mobile banking app.
    C. I can select the direct debit list, select a direct debit and see the full address of the subscription company.
    D. Login to the mobile banking app, select the menu option to see the list of direct debits, list appears within one second.

Question 37
What is a valid risk of test automation?

    A.   Relying on a tool too much.

    B.   Reduced execution times.

    C.   Developers change the code.

    D.   Testers have more time.

Question 38
Which of the following is the *most* important generic skill for testers?

    A.   Empathetic

    B.   Positive

    C.   Thorough

    D.   Decisive

Question 39
A vehicle insurance policy is subject to certain surcharges on the standard premium, added as percentages of the standard premium under certain conditions. The conditions are cumulative, so each condition that is met affects the premium.

The following decision table has been designed to test the logic for determining insurance premiums.

	T1	T2	T3	T4	T5	T6	T7	T8
Conditions								
3 penalty points	No	No	Yes	No	Yes	Yes	Yes	Yes
6+ penalty points	No	No	Yes	Yes	No	Yes	Yes	Yes
Age under 25	No	Yes	No	Yes	No	Yes	No	Yes
Action	Normal premium	+15%	+10%	+25%	+5%	+20%	+15%	+30%

Which test case must be eliminated because it is infeasible?

    A.   T1

    B.   T4

    C.   T5

    D.   T7

Question 40
Which of the following statements correctly describes the defect clustering testing principle?

A. Testing should be targeted at the most junior developer's code. This is where most defects will occur.

B. If no defects are found in the first 25 per cent of the time available, the code can be deemed safe to deliver to production.

C. Testing effort is to be targeted at what is thought to be the riskiest areas, and later those areas where there are more defects found.

D. Finding and fixing defects does not help if the system built is unstable and does not match user needs/expectations.

A2 MOCK CTFL EXAMINATION ANSWERS

Q1	B	Q11	A	Q21	C	Q31	A
Q2	C	Q12	D	Q22	B	Q32	C
Q3	D	Q13	A	Q23	D	Q33	B
Q4	A	Q14	C	Q24	D	Q34	A
Q5	A	Q15	A	Q25	B	Q35	B
Q6	B	Q16	A	Q26	B	Q36	A
Q7	B	Q17	B	Q27	B	Q37	A
Q8	A	Q18	C	Q28	D	Q38	C
Q9	C	Q19	B	Q29	B	Q39	B
Q10	A	Q20	C	Q30	D	Q40	C

A3 MOCK CTFL EXAMINATION COMMENTARY

Question 1
This is a K1 question relating to Learning Objective FL-1.2.2 – Recall the relation between testing and quality assurance.

Option A incorrectly describes testing as quality assurance; and quality assurance as testing.

Option C incorrectly suggests that quality assurance is used to fix defects, whereas it is designed to prevent defects.

Option D is an incorrect statement – testing and quality assurance are not the same.

The correct answer is Option B.

Question 2
This is a K1 question relating to Learning Objective FL-1.5.2 – Recall the advantages of the whole team approach.

Option A is not true and perhaps a more obvious option to eliminate.

Option B contradicts the syllabus, which states that the whole team approach is not always appropriate, in particular where high levels of test independence are required (e.g. for safety-critical projects).

Option D also contradicts the syllabus, which states that everyone is responsible for quality (not just the developer).

Option C is correct.

Question 3
This is a K2 question relating to Learning Objective FL-1.4.5 – Compare the different roles in testing.

Option A is not true, it does depend on factors such as project and product context.

Option B is incorrect as this is the responsibility of the testing role.

Option C is incorrect, the syllabus states that some management tasks may be handled by the Agile team, but it does not say that the management role is not required.

Option D is correct.

Question 4
This is a K1 question relating to Learning Objective FL-3.2.3 – Recall which responsibilities are assigned to the principal roles when performing reviews.

Option B incorrectly allocates the task of assigning staff to the review leader (this is a management function).

Option C incorrectly suggests that reviewers must be subject-matter experts, but reviewers can be drawn from any stakeholders as well as other specialists.

Option D incorrectly suggests that moderators decide who will be involved and the time/ place of a review (these are the review leader's responsibilities).

Option A correctly identifies the responsibilities of the manager role (Section 3.2.3).

Question 5
This is a K2 question relating to Learning Objective FL-2.2.1 – Distinguish the different test levels.

Option B – a database module – is incorrect; a database module could be a test object for component testing, but not a test basis.

Option C – an interface definition – is incorrect; this could be a test basis for integration testing.

Option D – a business process – is incorrect; this could be a test basis for acceptance testing.

Option A – a code specification – is correct, this is a suitable test basis for component testing.

Question 6
This is a K2 question relating to Learning Objective FL-1.2.1 – Exemplify why testing is necessary.

Testing cannot ensure that the right version of software is being delivered – that is the task of configuration management – so Option A is incorrect.

Testing in itself does not improve quality (Option C). Testing can identify defects, but it is in the fixing of defects that quality is actually improved.

Testing cannot show that software is error-free – it can only show the presence of defects (this is one of the seven testing principles), which rules out Option D.

Testing can be used to assess quality; for example, by measuring defect density or reliability, so Option B is correct.

Question 7
This is a K2 question relating to Learning Objective FL-3.1.2 – Explain the value of static testing.

Option A is a straightforward example of project delay; there is no indication of whether or not static testing was employed, so the value of static testing cannot be determined.

Option C indicates that test specifications were rigorously reviewed, but this did not prevent project overrun, so does not indicate the value of static testing.

Option D does not, in itself, demonstrate the value of static testing because it identifies a situation in which delay and overspend occurred, whether or not static testing was deployed.

Option B is the best option because it identifies a situation in which, although delay occurred early in the life cycle, because static testing was deployed the project still completed on time and on budget. This suggests that the use of static techniques may have made the development phase more efficient and effective.

Question 8
This is a K1 question relating to Learning Objective FL-2.1.3 – Recall the examples of test-first approaches to development

Options B, C and D are incorrect because these are all valid test-first approaches to development.

Option A is the correct answer because decision table-driven development (DTDD) is a made-up term and there is **not** a test-first approach to development.

Question 9
This is a K1 question relating to Learning Objective FL-5.3.1 – Recall metrics used for testing.

Option A is not a measure of testing, just a head count.

Option B is a count of how many test cases were written; it does not count test cases actually used in the tests.

Option D is a measure of the size of the test basis but not a metric of testing.

Option C is correct; it measures the total number of tests prepared and identifies how many were actually run and how many were not run.

Question 10
This is a K2 question relating to Learning Objective FL-2.1.1 – Explain the impact of the chosen SDLC on testing.

    i. is incorrect as it is not a direct impact on testing, it affects the whole project.

   ii. is correct.

  iii. is correct.

  iv. is correct.

   v. is incorrect as this will depend on many other factors, such as the size and scope of the change being delivered; it is **not** something directly impacted by the SDLC being followed.

Option A is therefore the correct answer – ii, iii and iv.

Question 11
This is a K1 question relating to Learning Objective FL-2.1.2 – Recall good testing practices that apply to all SDLCs

Options B, C and D are all good testing practices.

Option A is not a good testing practice; if anything it could be a good project management practice, albeit this is rarely true. The prioritisation of project funding will depend on many factors based on the business goals.

Question 12
This is a K2 question relating to Learning Objective FL-2.1.5 – Explain the shift-left approach.

Option A is incorrect as it also requires dynamic testing to be performed.

Option B is incorrect. Non-functional testing should be performed as early as possible.

Option C is incorrect. It should not increase the cost; if anything it should reduce it as fewer defects should be found in later testing.

Option D is correct.

Question 13
This is a K2 question relating to Learning Objective FL-4.1.1 – Distinguish black box, white box and experience-based test techniques.

1. Error guessing is an experience-based technique.
2. Equivalence partitioning is a black box technique.
3. Statement testing is a white box technique.
4. Boundary value analysis is a black box technique.
5. Branch testing is a white box technique.

Option A is therefore correct as it is the only option that correctly matches each technique to its correct type.

Question 14
This is a K2 question relating to Learning Objective FL-4.3.2 – Explain branch testing.

Option A is incorrect, this is the aim of statement testing.

Option B is incorrect, branch testing is typically performed during component testing and therefore early in the SDLC.

Option D is incorrect, it is possible to achieve 100 per cent coverage.

Option C is correct. A branch is a transfer of control between two nodes in the control flow graph, which shows the possible sequences in which source code statements are executed in the test object. Each transfer of control can be either unconditional (i.e. straight-line code) or conditional (i.e. a decision outcome).

Question 15
This is a K2 question relating to Learning Objective FL-4.3.3 – Explain the value of white box testing.

Options B, C and D are all correct.

Option A is the correct answer – this is not a value of white box testing; sometimes the code will not even be executable and even if it is, it is not generally complete enough to show stakeholders the system.

Question 16
This is a K2 question relating to Learning Objective FL-4.4.1 – Explain error guessing.

Option B is incorrect because, while such lists may be of value in avoiding defects, they are not used in error guessing.

Option C is incorrect because, while error guessing may be based partly on an individual tester's experience, the technique utilises other sources, such as failures that have occurred in other applications.

Option D is incorrect because this is an example of retesting, not error guessing.

Option A is correct and reflects the syllabus, Section 4.4.1.

Question 17
This is a K2 question relating to Learning Objective FL-5.1.3 – Compare and contrast entry criteria and exit criteria.

Option A is a project management issue and not an entry criterion for a testing phase.

Option C is not necessarily relevant, since some earlier testing phases may not have delivered components or subsystems for the part of the system about to be tested.

Option D is incorrect because not all defects reported in previous phases may be relevant, and clearance of defects that are relevant should be addressed by exit criteria from previous phases.

Option B is correct because testing cannot begin until a test environment is available.

Question 18
This is a K2 question relating to Learning Objective FL-3.2.2 – Summarise the activities of review process.

Option A has an initiation activity, a planning activity and the reviewing phase but then swaps the issue communication and analysis phase with the fixing and reporting phase.

Option B has a planning activity before the initiation phase, and also omits the individual review phase before moving on to issue communication and analysis and fixing defects phases.

Option D correctly includes items from the planning phase, the initiation phase and the individual review phase, but then moves straight to fixing defects before the details of defects have been communicated (part of 'issue communication and analysis').

Option C is correct because it includes one activity from each phase in the correct sequence.

Question 19
This is a K2 question relating to Learning Objective FL-4.5.2 – Classify the different options for writing acceptance criteria.

Option A is a rule-oriented format.

Option C is a rule-oriented format.

Option D is a user story format as per syllabus Section 4.5.1.

Option B is correct as it is a scenario-oriented format.

Question 20
This is a K2 question relating to Learning Objective FL-4.4.2 – Explain exploratory testing.

Option A is incorrect because exploratory testing does not use predefined tests.

Option B is incorrect because exploratory testing is not associated with model-based test strategies but may sometimes be associated with reactive test strategies.

Option D is incorrect because, while exploratory testing may sometimes use session-based testing, this is not characteristic of exploratory testing. When session-based testing is used, this is to structure the testing activity rather than to ensure that tests are documented.

Option C is correct and corresponds to a statement in Section 4.4.2 of the syllabus.

Question 21
This is a K2 question relating to Learning Objective FL-1.4.1 – Summarise the different test activities and tasks.

All of the options list activities from **one** of the groups of testing activities, so the task is to identify activities for test execution.

Option A is incorrect because it lists test planning activities.

Option B is incorrect because it lists test implementation activities.

Option D is incorrect because it lists test analysis activities.

Option C is correct because it lists test execution activities.

Beware of Option B – the activities listed are **preparation** for test execution, rather than for the running of tests themselves.

Question 22
This is a K2 question relating to Learning Objective FL-5.1.7 – Summarise the testing quadrants and their relationships with test levels and test types.

The test types listed are taken directly from the syllabus Section 5.1.7.

Option B is the correct answer; it is the only option that matches all types to the correct quadrants.

Question 23
This is a K3 question relating to Learning Objective FL-5.1.4 – Use estimation techniques to calculate the required test effort.

Option A is incorrect as it is $(a + m) / 2$.

Option B is incorrect and would not be arrived at through any formula.

Option C is incorrect as it is just $=b$.

Option D is correct as it uses the most popular version of this technique:
$(a + 4 \times m + b) / 6$, so $10 + 80 + 30 = 120$; then divided by $6 = 20$.

Question 24
This is a K2 question relating to Learning Objective FL-6.1.1 – Explain how different types of test tools support testing.

Option A is incorrect because static analysis tools support the tester in performing reviews and static analysis.

Option B is incorrect because test-driven development tools facilitate generation of test cases, test data and test procedures.

Option C is incorrect because collaboration tools facilitate communication.

Option D is the correct answer because coverage tools facilitate automated test execution and coverage measurement.

Question 25
This is a K2 question relating to Learning Objective FL-5.4.1 – Summarise how configuration management supports testing.

Option A is correct but incomplete because it mentions only the testware and system components separately, and there is no mention of the relationships between items.

Option C is correct but incomplete because it relates only to documentation.

Option D is incorrect because it relates only to test items.

Option B correctly refers to managing the system components, the testware and the relationship between them.

Question 26
This is a K3 question relating to Learning Objective FL-4.2.1 – Use equivalence partitioning to derive test cases.

The required partitions are:

- >£10,000
- £8,001–£10,000
- £5,001–£8,000
- £3,001–£5,000
- <£3,001

There are five partitions, so Option B is correct.

Question 27
This is a K2 question relating to Learning Objective FL-1.4.4 – Explain the value of maintaining traceability.

Option A is incorrect. This could help in clarifying whether all requirements are covered by one or more test cases, but the requirements could be included in a test case that was not actually run.

Option C is incorrect because traceability has no direct bearing on whether the project will be delivered on time, and it is not something that is stated (in the syllabus or elsewhere) as a 'benefit' of traceability.

Option D is incorrect and inappropriate in that it implies a 'blame culture' rather than one where cooperation and product quality are to the forefront.

Option B is correct and is specifically mentioned in Section 1.4.4 of the syllabus.

Question 28
This is a K3 question relating to Learning Objective FL-4.2.2 – Use boundary value analysis to derive test cases.

Option A is incorrect because it starts each time zone on the boundary rather than just under, but ends each time zone correctly.

Option B is incorrect because it starts each time zone correctly but does not check the end of each time zone correctly.

Option C is incorrect because it starts each time zone on the boundary rather than before the boundary and ends each time zone too early.

Option D is correct because it correctly tests the values just before and on the lower boundaries, and on and just over the higher boundaries.

Question 29
This is a K1 question relating to Learning Objective FL-3.2.5 – Recall the factors that contribute to a successful review.

Option A is incorrect: it is always important for users to review requirements documents; the presence of designers, developers and testers may help with any technical terms.

Option C is incorrect: an inspection is not likely to be appropriate for this document and one key success factor is that an appropriate review type is applied. Metrics are not important at this stage, but removal of ambiguity, clarity of expression and understandability for users are vital.

Option D is incorrect: reviews should always be scheduled with adequate notice and time for participants to prepare.

Option B is the correct answer (in line with Section 3.2.5 in the syllabus): it will allow reviews to begin earlier and will provide earlier feedback to authors to enable improvements to be made continually.

Question 30
This is a K2 question relating to Learning Objective FL-2.3.1 – Summarise maintenance testing and its triggers.

Option A is incorrect because a new feature required for an iteration implies development rather than maintenance.

Option B is incorrect because it again implies development rather than maintenance.

Option C is incorrect for the same reason.

Option D is correct because data migration is typically a maintenance activity.

Question 31
This is a K2 question relating to Learning Objective FL-5.2.3 – Describe, by using examples, how product risk analysis may influence the thoroughness and scope of testing.

Options B, C and D are incorrect because they are all related to managing the project.

Option A specifically relates to the testing activities and how these need to be driven by risk levels.

Question 32
This is a K2 question relating to Learning Objective FL-5.3.3 – Exemplify how to communicate the status of testing.

Option A is incorrect: one or more options can be used.

Option B is incorrect and contradicts the syllabus, which states that more formal communication may be more appropriate for distributed teams.

Option D is incorrect: verbal communication is listed as an option in the syllabus and is one of the most common methods used within a team.

Option C is correct: 'Typically, different stakeholders are interested in different types of information, so communication should be tailored accordingly.'

Question 33
This is a K3 question relating to Learning Objective FL-5.1.5 – Apply test case prioritisation.

Option A was the original schedule, but this has now been altered, so Option A is incorrect.

Option C places test cases 5, 3 and 2 in the correct sequence but test case 6, which is high-priority, is relegated to fifth place, so Option C is incorrect.

Option D places 6 in second position but does not make test case 3 dependent on test case 5, so Option D is incorrect.

Option B is the correct answer because it correctly sequences 5, 3 and 2 and places test case 6 ahead of this trio.

Question 34
This is a K3 question relating to Learning Objective FL-4.2.4 – Use state transition testing to derive test cases.

Option B is incorrect because an invalid transition is represented (test case 6).

Option C is incorrect because all of the valid transitions are represented (test cases 1–5 correspond to the five valid transitions shown in the diagram).

Option D is incorrect because all valid transitions are represented, and one invalid transition is addressed in test case 6.

Option A is correct because the table correctly identifies the five valid transitions (test cases 1–5) and one invalid transition (test case 6).

Question 35
This is a K3 question relating to Learning Objective FL-5.5.1 – Prepare a defect report.

Options A, C and D are all incorrect because they list valid fields on a defect report, but none of them prevents action being taken to resolve the problem.

Option B is the correct answer because the actual result is needed to enable corrective action to be correctly applied.

Question 36
This is a K3 question relating to Learning Objective FL-4.5.3 – Use acceptance test-driven development (ATDD) to derive test cases.

Option B is incorrect as this is another user story, not a test case.

Option C is incorrect as this is an acceptance criterion, not a test case.

Option D is incorrect as this is a non-functional test, not a positive test.

Option A is correct as this is a positive test case as mentioned in Section 4.5.3 of the syllabus: 'confirming the correct behaviour without exceptions or error conditions, and comprising the sequence of activities executed if everything goes as expected'.

Question 37
This is a K1 question relating to Learning Objective FL-6.2.1 – Recall the benefits and risks of test automation.

Option B is incorrect: this is actually a benefit of automation.

Option C is incorrect: while this may happen, it is not really a recognisable risk and certainly not stated in the syllabus.

Option D is incorrect: this is also a benefit of test automation.

Option A is correct: it is a risk and stated in Section 6.2 of the syllabus, 'Relying on a tool, when human critical thinking is what is needed'.

Question 38
This is a K2 question relating to Learning Objective FL-1.5.1 – Give examples of the generic skills required for testing.

Options A, B and D are all good skills for anyone to have, but not particularly relevant for testers.

Option C is correct as thoroughness is clearly stated in the syllabus as a key skill.

Question 39
This is a K3 question relating to Learning Objective FL-4.2.3 – Use decision table testing to derive test cases.

Option A is feasible and will apply to most applicants, so it is not the correct answer.

Options C and D both involve penalising for 3 points and for 6+ points, which are valid within the rules, so neither of these is the correct answer.

Option B is infeasible because it is not possible to incur 6 penalty points without also incurring 3 penalty points.

Option B is therefore the correct answer.

Be careful in questions like this one to note that the question asked for **infeasible** test cases. It is easy, especially under time pressure, to opt for the more usual expectation of identifying feasible test cases.

Question 40
This is a K2 question relating to Learning Objective FL-1.3.1 – Explain the seven testing principles.

Option A appears to reflect the defect clustering principle, but there is no reason to assume that defect clustering will be associated with the work of the most junior developer, so Option A is incorrect.

Option B is an example of the 'absence of errors' fallacy – the fact that no errors have been found does not mean that code can be released into production – so Option B is incorrect.

Option D is a direct statement of the 'absence of errors' fallacy, so Option D is incorrect.

Option C is the best statement of the defect clustering principle – defects are often found in the same places, so testing should focus first on areas where defects are expected or where defect density is high. Option C is therefore the correct answer.

INDEX